Graded Exercises in Electrical and Electronic Engineering

Christopher R Robertson

Formerly Head of School
West Kent College
Tonbridge
Kent

ARNOLD

A member of the Hodder Headline Group
LONDON • SYDNEY • AUCKLAND

621·387 R

First published in Great Britain 1996 by
Arnold, a member of the Hodder Headline Group,
338 Euston Road, London NW1 3BH

British Library Cataloguing in Publication Data
A catalogue record for this book is available from the British Library

ISBN 0 340 64566 0

Typeset in 10/12pt Times and produced by Gray Publishing, Tunbridge Wells, Kent
Printed and bound in Great Britain by J. W. Arrowsmith Ltd, Bristol

Contents

Introduction

This book is designed to complement the two volumes *Electrical and Electronic Principles 1* and *2*. Due to the graded nature of the assignment questions, many of them are quite demanding, and will therefore also be found of use for Higher National, first-year undergraduate studies in electrical engineering, and associated bridging courses. Of necessity, the assignment questions at the end of each chapter of most textbooks tend to concentrate solely on the topic covered by the relevant chapter. However, this tends to fragment the subject matter. Consequently the student, once tested, tends to 'forget' about earlier topics and concentrates solely on the current topic of study. This effect is compounded by the current system of phase tests and assignments in preference to a comprehensive end test on completion of the unit of study.

The objective of this book is to present more realistic engineering problems. In many cases this means that the student has to utilise knowledge gained over a range of topics in order to arrive at a solution. This will help the student to view the unit(s) as a cohesive whole, rather than isolated pockets of knowledge. In order to enhance the integrative aspect, some exercises include topics from the BTEC Electronics syllabuses together with some elements from the Electrical Applications. The subject matter of this last unit has considerable overlap with that of *Electrical and Electronic Principles*.

Very broadly, Chapters 1–4 relate to the subject matter studied during the first year of a BTEC course. The remaining chapters relate to the areas of study in *both* the second and first years. In order to overcome undue cross-referencing between books the relevant equations and constants are listed at the beginning of each chapter.

Worked examples are not included since there is a wealth of examples in the two companion volumes. The numerical answers to the exercises are given at the end of the book.

C. R. Robertson
Tonbridge, Kent

1 Direct Current Circuits

Equations

❏ **Note:** Whenever numerical values for quantities are inserted into an equation, then these quantities *must* be expressed in their basic units. For example, length in metres; area in square metres; time in seconds; current in amps, etc.

Application of Ohm's law

$$V = IR \text{ volt}$$

Affect of dimensions on resistance

$$R = \frac{\rho l}{A} \text{ ohm}$$

Affect of temperature on resistance

$$R_1 = R_0(1 + \alpha\theta_1) \text{ ohm}$$

$$\text{or } R_2 = R_1\left(\frac{1+\alpha\theta_2}{1+\alpha\theta_1}\right) \text{ ohm}$$

where R_0 = resistance at 0°C, R_1 = resistance at θ_1°C, R_2 = resistance at θ_2°C and α = temperature coefficient of resistance.

Energy and power

$$W = VIt = I^2Rt = \frac{V^2}{R}t \text{ joule}$$

$$P = VI = I^2R = \frac{V^2}{R} \text{ watt}$$

❏ **Note:** The commercial 'unit' of energy is the kilowatt hour (kWh).

Resistors in series

$$R = R_1 + R_2 + \ldots + R_n \text{ ohm}$$

for n resistors in series.

Resistors in parallel

$$\frac{1}{R} = \frac{1}{R_1} + \frac{1}{R_2} + \ldots + \frac{1}{R_n} \text{ siemen}$$

or $G = G_1 + G_2 + \ldots + G_n$ siemen

where G is the conductance of the circuit, i.e $G = 1/R$ siemen

$$R = \frac{R_1 R_2}{R_1 + R_2} \text{ ohm}$$

when *only two* resistors are in parallel.

Terminal potential difference (p.d.) of an electromotive force (e.m.f.) source

$$V = E - Ir \text{ volt}$$

where r is the internal resistance of the source.

Potential divider

$$V_1 = \frac{R_1}{R_1 + R_2} E \text{ volt}$$

$$V_2 = \frac{R_2}{R_1 + R_2} E \text{ volt}$$

Current divider

$$I_1 = \frac{R_2}{R_1 + R_2} I \text{ amp}$$

$$I_2 = \frac{R_1}{R_1 + R_2} I \text{ amp}$$

Kirchhoff's laws

$$\Sigma I = 0$$

or $I_1 + I_4 = I_2 + I_3$ amp

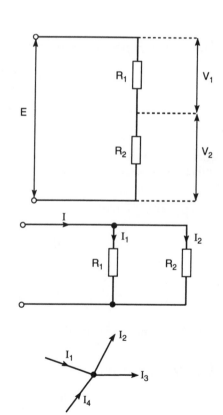

$\sum E = \sum IR$ volt

or $E_1 - E_2 = I_1R_1 - I_2R_2$ volt, etc.

Constants

Resistivity ρ (rho) and temperature coefficient of resistance α for some common materials.

Material	ρ (Ω m)	α (/°C)
Aluminium	2.65×10^{-8}	40×10^{-4}
Brass	8×10^{-8}	15×10^{-4}
Constantan	47×10^{-8}	0.4×10^{-4}
Copper	1.7×10^{-8}	39×10^{-4}
Gold	2.4×10^{-8}	34×10^{-4}
Iron	10×10^{-8}	65×10^{-4}
Phosphor bronze	7×10^{-8}	60×10^{-4}
Silver	1.6×10^{-8}	40×10^{-4}

Assignment Questions

1.1 One of the standard resistors in a wheatstone bridge instrument is to be made from a length of constantan wire of diameter 0.7 mm. Determine the total length of wire required to form a resistance of 1 Ω.

1.2 A length of copper wire has a resistance of 2.4 Ω at 0°C. Determine its resistance when at a temperature of 30°C.

1.3 The copper field coil of an electric motor has a resistance of 200 Ω at a temperature of 18°C. Under normal operating conditions the temperature of this coil increases to 32°C. Determine
   (a) the current drawn by this coil from a 400 V d.c. supply under normal operating conditions, and
   (b) the winding resistance at 0°C.

1.4 An electrical load is supplied with a current of 50 A via an aluminium and a copper cable which are connected in parallel. The length of each cable is 350 m and each has a cross-sectional area of 40 mm². Determine
   (a) the voltage drop along the length of the combined cables
   (b) the current carried by each cable, and
   (c) the power wastage in each cable.

1.5 A portable compact disk player normally operates from a 9 V d.c. supply, drawing a current of 250 mA. It is required to be able to use the player in a car by plugging it into the 12 V cigar lighter socket. Determine the value and power rating of the resistor that must be connected in series with the player to enable normal operation.

1.6 A length of copper wire is to be formed by drawing out a copper rod. The rod is 0.5 m long, 5 mm in diameter, and has a resistance of 450 $\mu\Omega$ at 18°C.
(a) Calculate the resistivity of the copper at 18°C.
(b) If the drawn wire has a uniform diameter of 0.6 mm calculate its resistance at both 18°C and 50°C. You may assume that the resistivity remains unchanged.

1.7 A moving coil meter uses two contrawound phosphor bronze restoring springs. The deflecting current passes through the coil via these springs (effectively in series). Each spring has a total length of 10 cm, being 2 mm wide and 0.5 mm thick. Determine the voltage drop due to the springs when the current through the coil is 100 μA.

1.8 A 12 V d.c. supply of internal resistance 0.2 Ω is used to supply an emergency lighting circuit. Each of the light bulbs has a resistance of 6 Ω and requires a minimum p.d. of 9 V in order to operate correctly. Calculate how many of these bulbs may be satisfactorily connected in parallel with the above supply.

1.9 A 3 kW electric kettle is designed to operate with a 240 V supply. Calculate the current drawn and energy dissipated, in kilowatt-hours, when it is operated from a 220 V supply for a period of 6 minutes.

1.10 A piece of laboratory equipment operated from a 24 V d.c. supply has an internal potential divider circuit as in Fig. 1.1. The open-circuit p.d. between terminals A and B is required to be 6.5 V ± 5%. The following resistor values are available in power ratings of 1/8 W, 1/4 W and 1/2 W, all with a resistance tolerance of ±5%: 390 Ω; 470 Ω; 560 Ω; 750 Ω; and 1 kΩ.
(a) Determine
 (i) the resistor value and power rating suitable for R_2, allowing for the full resistance tolerance. You may assume that R_1 is exactly the value shown.
 (ii) The appropriate power rating for R_1.

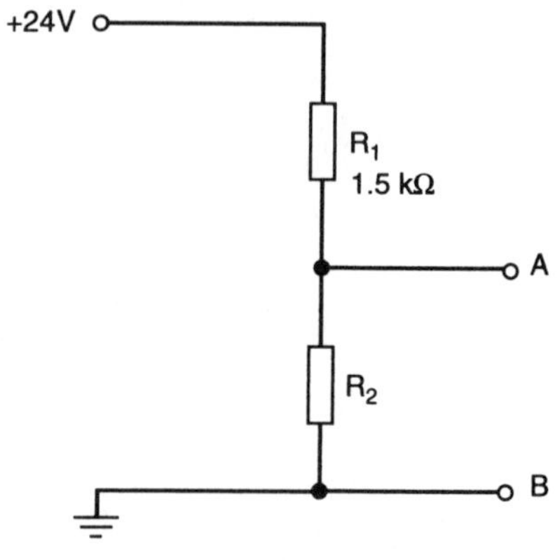

Fig. 1.1.

(b) If a load of effective resistance 7.5 kΩ is connected between terminals A and B, calculate the voltage now available between the terminals, and check whether this is still within the required tolerance. In this case you may assume that both resistors are of their nominal value.

1.11 A portable d.c. generator has an e.m.f. of 110 V and internal resistance of 0.05 Ω. It is to be used with a twin-core cable 250 m in length. The system specification requires that the available p.d. at the far end of the cable must be not less than 100 V when supplying a current of 45 A. Two types of twin-core cable are available: one having copper conductors each of cross-section 40 mm^2; the other having aluminium conductors, each of cross-section 80 mm^2. Determine
(a) which of the two cables will meet the required specification, and
(b) the maximum current that can be supplied by the chosen cable whilst still meeting the criterion of 100 V p.d. available at the load.

1.12 Part of a circuit requires a resistor of 1.02 kΩ. However, this precise value is not directly available, and has to be formed by connecting two or more resistors in a series/parallel combination. From the following list of available resistors: 150 Ω, 270 Ω, 390 Ω, 470 Ω, 560 Ω, 1 kΩ, 1.2 kΩ, 1.5 kΩ, 2.2 kΩ

   (a) devise a circuit to meet the requirement (hint – you may use more than one of each resistor if required), and

   (b) determine the power dissipation of each resistor used when a p.d. of 15 V exists across the combined circuit.

1.13 The battery in a vehicle has an e.m.f. of 12 V and at 20°C its internal resistance is 0.01 Ω. When the starter motor is operated it draws a current of 150 A. The ignition coil must be supplied with a minimum of 9 V in order for the engine to start. Given that the battery electrolyte has a temperature coefficient of resistance of -16.7×10^{-3}/°C (i.e. a **negative** coefficient), determine the minimum temperature at which the engine will start.

1.14 The resistance of the coil in a moving coil meter is 40 Ω. In order to provide full-scale deflection the coil current required is 250 μA. Determine the value of shunt resistor to be connected in parallel with the coil so that full-scale deflection is obtained when the total current drawn by the meter is 5 A.

1.15 A voltage attenuator network feeding a 500 Ω load (R_L) is as shown in Fig. 1.2. Calculate the effective resistance measured between terminals A and B.

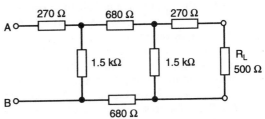

Fig. 1.2.

1.16 The d.c. biasing for a transistor amplifier is to be achieved by resistors R_1 and R_2 shown in Fig. 1.3. The transistor base current (I_B) is 200 μA, and current I_1 must be approximately $20 \times I_B$. It is also required that the p.d. across R_2 be 1.3 V.

   (a) Determine suitable values for R_1 and R_2 and hence select the appropriate values from the following list of preferred values: 270 Ω, 330 Ω, 390 Ω, 470 Ω, 1.5 kΩ, 1.8 kΩ, 2.2 kΩ, 2.7 kΩ.

   (b) For the resistors selected, determine the ratio of I_1 to I_B actually achieved, assuming I_B remains at 200 μA.

   (c) For the two resistors selected calculate the power dissipation in each.

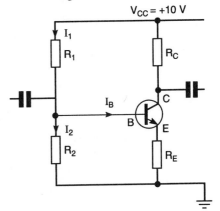

Fig. 1.3.

1.17 Two batteries are connected in parallel with each other and this combination is connected across a 15 V charging source of internal resistance 0.1 Ω. Given that the initial e.m.f. and internal resistance of one battery is 10.5 V and 0.2 Ω; the corresponding values for the second battery being 10 V and 0.4 Ω, then calculate (a) the initial charging current into each battery, and (b) the energy absorbed by each battery over the first five minutes of charging, if the initial charging currents remain constant for this period of time.

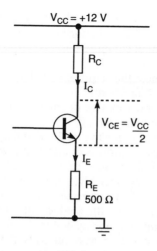

Fig. 1.4.

1.18 Part of the d.c. biasing circuit for a transistor is as shown in Fig. 1.4. Given that $I_C = 2$ mA and assuming that for practical purposes $I_C = I_E$, determine
(a) the p.d. across R_C
(b) the value of R_C, and
(c) the powers dissipated by R_C and R_E.

1.19 A wheatstone bridge resistance measuring instrument is illustrated in Fig. 1.5. The current detector (M) has a resistance of 100 Ω. Determine
(a) the value to which R_{BC} must be set in order to balance the bridge
(b) the potentials at points B and D under this condition, and
(c) the current flowing through M when R_{BC} is set to 1.8 kΩ.

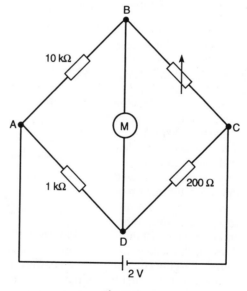

Fig. 1.5.

1.20 Figure 1.6 shows a wire resistance strain gauge bridge network, where R_1 to R_4 represent the strain gauges. Gauges R_1 and R_3 are attached to a cantilever beam which is to be subjected to a bending force. The normal unstrained resistance of each gauge is 120 Ω. When the force is applied, the resistance of R_1 increases by 0.2% whilst that of R_3 decreases by the same percentage. An instrumentation amplifier is connected between terminals A and B of the bridge. This amplifier has such a high input resistance that it draws negligible current from the bridge circuit. Calculate the value and polarity of the p.d. applied to this amplifier when the force is applied to the beam.

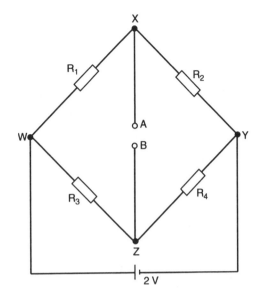

Fig. 1.6.

1.21 A temperature measuring system utilises a thermistor (a temperature-dependent resistor) as one of the arms of a wheatstone bridge circuit, as illustrated in Fig. 1.7. The resistance/temperature characteristics of the thermistor (R_4 in the circuit) are shown in the graph of Fig. 1.8 on page 8. The circuit is to be initially calibrated so that when the thermistor is subjected to a temperature of 15°C, R_3 is adjusted to balance the bridge. (The detector D indicates zero.) Calculate

(a) the value to which R_3 must be preset in order to perform the initial calibration

(b) the p.d. applied to the detector when the thermistor temperature is 70°C if
  (i) the detector has an infinite input resistance, and
  (ii) the detector has an input resistance of 20 kΩ.

You may assume that the values of R_1 to R_3 remain unchanged.

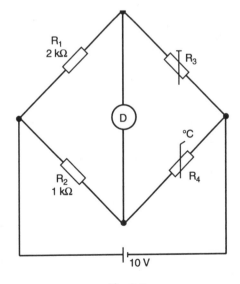

Fig. 1.7.

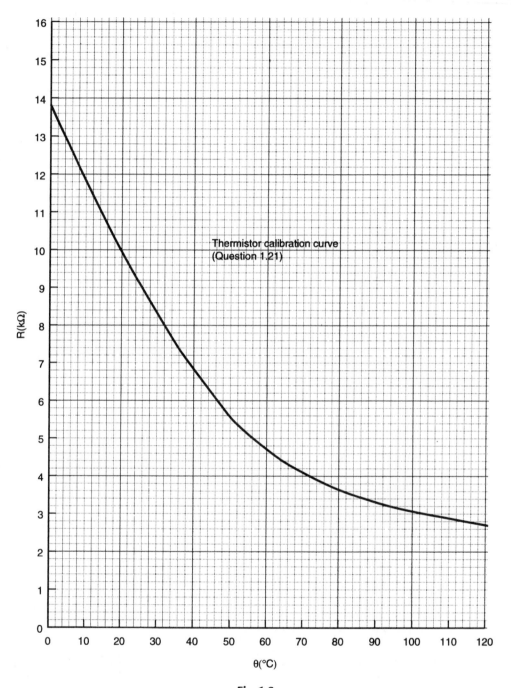

Thermistor calibration curve
(Question 1.21)

Fig. 1.8.

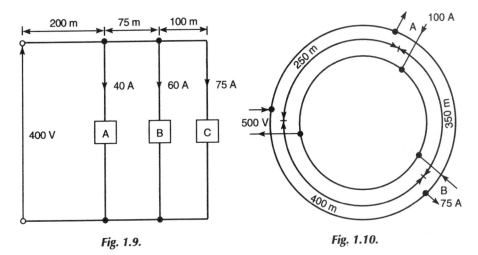

Fig. 1.9.   **Fig. 1.10.**

1.22 A power distribution system consists of two feeder lines supplied at 400 V, supplying three consumers, A, B and C at various points along the system, as shown in Fig. 1.9. Consumer A draws a current of 40 A; the corresponding figures for customers B and C being 60 A and 75 A, respectively. The resistance of the feeder lines is 0.01 Ω per 100 m of single line. Determine
(a) the p.d. available for each consumer, and
(b) the total power loss in the distribution system.

1.23 A d.c. ring distributor of effective length 1 km is supplied at 500 V. At points B and C are consumers drawing currents of 100 and 75 A, respectively, as shown in Fig. 1.10. Calculate the following
(a) the current in each section of the distributor (hint – use Kirchhoff's laws around one closed loop), and
(b) the p.d.s available at points B and C, given that the resistance per 100 m of single conductor is 0.02 Ω.

1.24 The equivalent circuit for a transistor amplifier is shown in Fig. 1.11. Determine the p.d. (v_2) across and the power developed in the 2 kΩ load resistor (R_L).

1.25 A rotary potentiometer is used as a rotary displacement transducer. The resistance element covers a total of 340°, has a total resistance of 10 kΩ, and is supplied at 15 V as shown in Fig. 1.12 overleaf. Determine

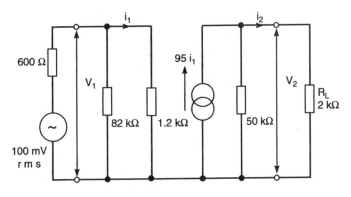

Fig. 1.11.

(a) the transducer transfer function, in volt/rad
(b) the ideal output voltage for a displacement of 2.5 rad from the zero position
(c) the actual voltage output at this displacement if a voltmeter, using its 10 V range, and having a figure of merit of 20 kΩ/V is connected across the output terminals, and
(d) the percentage voltage error due to the loading effect of the meter.

1.26 (a) Explain the terms zener voltage and diode slope resistance as applied to a zener diode.

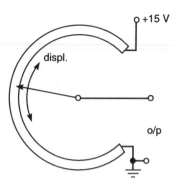

Fig. 1.12.

(b) A d.c. voltage of 15 V ± 5% is required to be supplied from a 24 V unstabilised source. It is proposed to achieve this by means of a simple voltage regulator circuit comprising a zener diode and protection resistor as shown in Fig. 1.13. The available resistors and zener diodes are listed below.

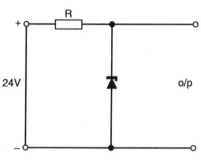

Fig. 1.13.

(i) For each diode listed, perform the necessary calculations in order to choose the appropriate protection resistor and its power rating. Hence determine the component cost for each circuit.

(ii) Assuming that the output voltage specification of 15 V ± 5% is achievable with a variation in zener current of 30 mA, determine which of the three circuits meets the specification at the lowest cost.

Diode No.	V_z (V)	Slope resistance (Ω)	Max. power (W)	Unit cost (£)
1	15	30	0.5	0.07
2	15	15	1.3	0.20
3	15	2.5	5.0	0.67

Resistors available in the following values (Ω):
18; 27; 56; 100; 120; 150; 180; 150; 220; 270; 330.

0.25 W resistors	£0.026 each
0.5 W resistors	£0.038 each
1.0 W resistors	£0.055 each
2.5 W resistors	£0.26 each
7.0 W resistors	£0.28 each

1.27 When the output from a logic gate circuit goes to the logic 1 state (+5 V) it is required to energise a relay. However, since the power output of the logic circuit is insufficient to operate the relay directly, it is proposed to use a transistor as the switching element. The circuit arrangement is shown in Fig. 1.14, where R_C represents the relay coil. The transistor data are listed below, together with the resistors available. Determine the appropriate resistor value for R_B.

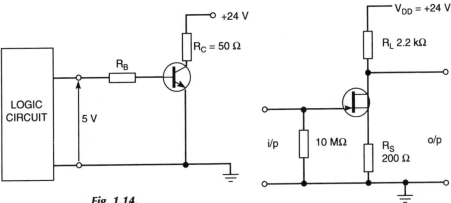

Fig. 1.14.

Fig. 1.15.

Transistor data:

$V_{BE(sat)} = 0.7$ V

$V_{CE(sat)} = 0.2$ V

$h_{FE} = 200$ (min) to 300 (max)

Resistor values: 2.7 kΩ; 3.3 kΩ; 3.9 kΩ; 4.7 kΩ; 5.6 kΩ; 6.8 kΩ; 8.2 kΩ.

1.28 The FET (field effect transistor) used in the amplifier circuit of Fig. 1.15 has characteristics as given in Table 1.1 above. Using this data

(a) plot the FET output characteristics, and on this graph plot the d.c. load line

(b) using the plotted load line determine a suitable operating (Q) point and state the resulting quiescent values for V_{GS}, I_D and V_{DS}

(c) determine the transistor drain-source resistance (R_{DS}) and the mutual conductance (G_m) under quiescent conditions, and

(d) the amplifier voltage gain when an input signal causes V_{GS} to vary by ±1 V about its quiescent value.

Table 1.1

| V_{DS} (V) | I_D (mA) | | | | |
	$V_{GS} = -2.5$ V	$V_{GS} = -2.0$ V	$V_{GS} = -1.5$ V	$V_{GS} = -1.0$ V	$V_{GS} = -0.5$ V
4	1.4	2.8	4.4	6.1	8.0
16	1.6	3.0	4.6	6.3	8.2
24	1.8	3.1	4.7	6.4	8.4

1.29 The transistor used in the circuit of Fig. 1.16 has input and output characteristics as shown in Figs 1.17 and 1.18, respectively, overleaf. Using these characteristics determine the following. Where appropriate, choose the resistor with the nearest value from the list at the end of this question

(a) the transistor input resistance $R_{in}(h_{ie})$

(b) a suitable value for R_C, given that under quiescent conditions (no input signal) $V_{CE} \approx V_{CC}/2$ volt and $V_{BE} = 0.7$ V

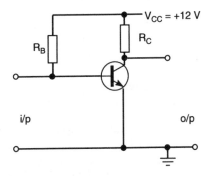

Fig. 1.16.

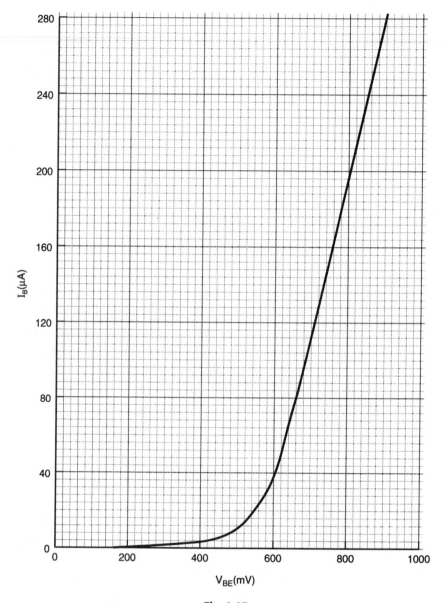

Fig. 1.17.

(c) the resulting quiescent values for V_{CE}, I_C, and I_B
(d) a suitable value for bias resistor R_B
(e) the *transistor* current gain, β (h_{FE}) under quiescent conditions
(f) the transistor output resistance, R_{out} ($1/h_{oe}$) under quiescent conditions
(g) the *amplifier* current and voltage gains when an input signal causes V_{BE} to
   vary by ± 100 mV about its quiescent value.

Resistor values available: 390 Ω; 470 Ω; 560 Ω; 680 Ω; 820 Ω; 1 kΩ; 1.2 kΩ;
68 kΩ; 82 kΩ; 100 kΩ; 120 kΩ; 150 kΩ.

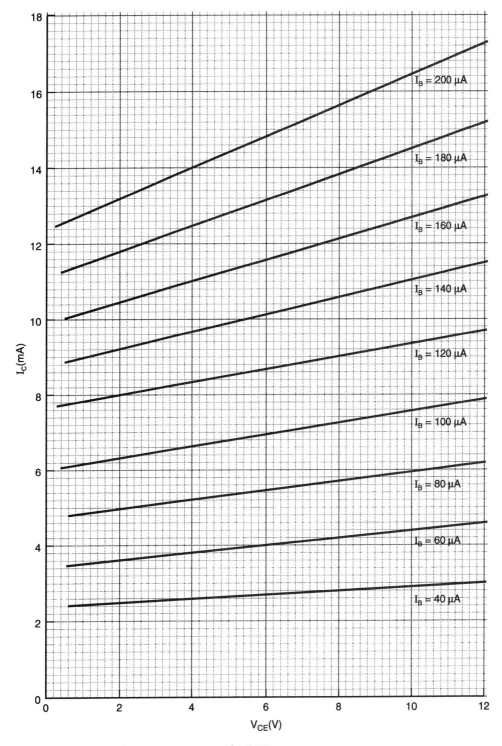

Fig. 1.18.

1.30 The total connected load of an electrical distribution system is 18 MW and the peak load is 11.25 MW. Calculate the load and diversity factors if the energy transmitted over a 120-hour period is 720 MWh.

1.31 A temperature warning circuit is shown in Fig. 1.19. In the unoperated state the relay contacts are open, but must close when the temperature reaches 85°C in order to activate the device. The relay rating is 12 V, 400 Ω and its operating voltage range is 8 to 12 V. The characteristics for the transistor and thermistor are given in Figs 1.20 (opposite) and 1.8, respectively.
  (a) determine the minimum relay coil current needed to operate the relay
  (b) insert the relevant load line on the transistor output characteristics and hence estimate the minimum base current required to cause the relay to operate
  (c) determine the corresponding base potential, and
  (d) calculate the required value for the resistor *R* shown in Fig. 1.19.

1.32 A light-dependent resistor (LDR) is to be used in the measurement of the illumination (in lux) onto a surface, the circuit arrangement being as shown in Fig. 1.21 below. The scale of the 0–1 mA ammeter is to be calibrated in lux, such that the full-scale deflection corresponds to an illumination of 1000 lux. The meter has a resistance of 150 Ω and the resistor *R* in the diagram is a preset resistor to enable calibration. With the aid of the resistance/illumination characteristics in Fig. 1.22 (shown on page 16), calculate
  (a) the value to which *R* must be set
  (b) the illumination that will produce half full-scale deflection, and
  (c) the meter deflection corresponding to an illumination of 90 lux.

1.33 A d.c. two-wire distribution system is shown in Fig. 1.23 on page 17, where the parallel loads are indicated by the currents drawn by each load.
  (a) calculate the voltages available at points B, C, D, and E given that the resistance of each conductor is 0.4 Ω/1000 m
  (b) if 240 V was also supplied at end E, in opposition to that at A, what would then be the voltages at points B, C, and D?
  (c) how can a system such as in (b) be achieved in practice, without the use of a second generator, and what are the resulting advantages?

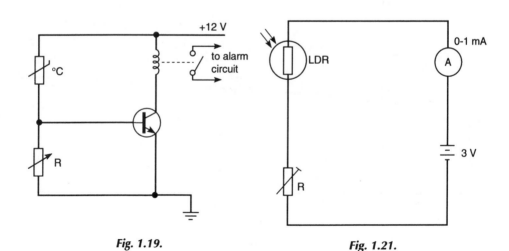

Fig. 1.19.   *Fig. 1.21.*

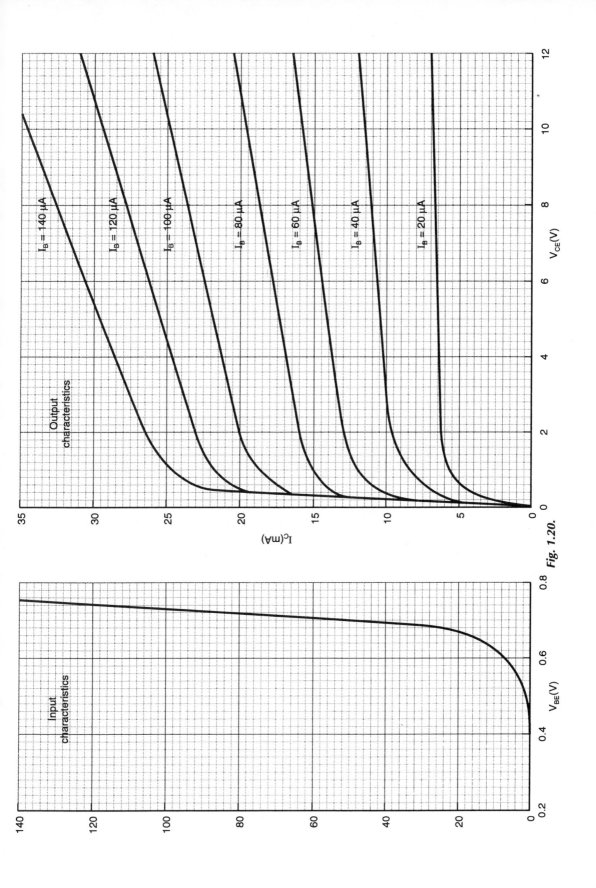

Fig. 1.20.

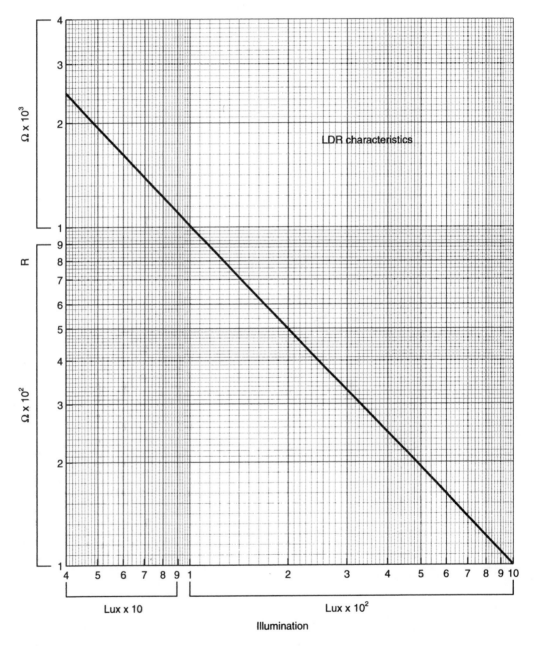

Fig. 1.22.

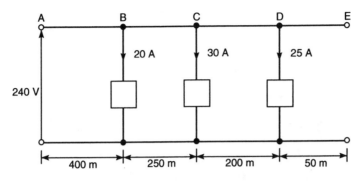

Fig. 1.23.

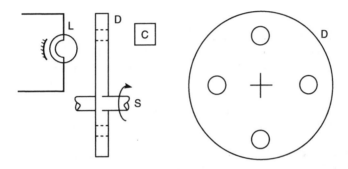

Fig. 1.24.

1.34 Figure 1.24 above illustrates a system for measuring the angular velocity of a shaft (S) by means of a disc (D) having four equally spaced holes; a photovoltaic cell (C) and a lamp (L). Ambient light is excluded from the cell by means of a hood (not shown), so that illumination from the lamp only will excite it whenever a hole in the disk is in the appropriate position. The time response of the cell to incident light is shown in Fig. 1.25, and it may be assumed to have this response regardless of the speed of the shaft.

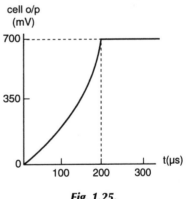

Fig. 1.25.

Figure 1.26 on the next page shows the circuit arrangement. The transistor becomes saturated (acts as a closed switch) when the potential applied to its base terminal is 0.7 V, the switching action taking a time of 50 μs. A monostable circuit (M) is triggered at the instant that its input falls to zero.

At this time its output rises linearly from zero to +5 V in 45 μs, remains at +5 V for a further 910 μs, and then falls linearly to zero in another 45 μs. In order to avoid spurious operation of the circuit, the end of one pulse of monostable output V_o must occur before light again falls onto the photovoltaic cell (C)

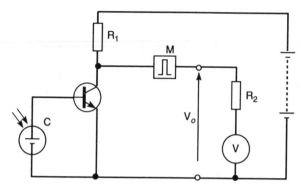

Fig. 1.26.

(a) calculate the maximum speed of rotation (in rev/min) of shaft S that can be measured without causing spurious operation of the circuit

(b) sketch the waveform of V_o over a period of 5 ms at maximum shaft speed, indicating the principal values of time and voltage. You may assume that time $t = 0$ corresponds to the instant when light from the lamp first falls onto the cell

(c) calculate the average value of V_o corresponding to the maximum shaft speed

(d) if the moving coil voltmeter (V) has a figure of merit of 10 kΩ/V and a coil resistance of 750 Ω, determine a suitable value for resistor R_2 such that maximum shaft speed results in full-scale deflection.

2 Electrostatics

Further exercises involving capacitors (e.g. time constant, etc.) are to be found in Chapter 5.

Equations

Charge or quantity of electricity

$$Q = It \text{ coulomb}$$

$$\text{or } Q = VC \text{ coulomb}$$

Force between charged bodies

$$F = \frac{Q_1 Q_2}{\varepsilon_0 \varepsilon_r d^2} \text{ newton}$$

Electric flux density

❏ **Note:** The electric flux has the same numerical value, the same quantity symbol and same unit of measurement as electric charge, Q

$$D = \frac{Q}{A} \text{ coulomb/metre}^2$$

$$\text{or } D = \varepsilon_0 \varepsilon_r E \text{ coulomb/metre}^2$$

Potential gradient, or electric field strength

$$E = \frac{V}{d} \text{ volt/metre}$$

$$\text{or } E = \frac{F}{q} \text{ newton/coulomb}$$

Capacitance

$$C = \frac{\varepsilon_0 \varepsilon_r A(n-1)}{d} \text{ farad}$$

where n = number of parallel plates.

Capacitors in series

$$\frac{1}{C} = \frac{1}{C_1} + \frac{1}{C_2} + \ldots + \frac{1}{C_n}$$

for n capacitors in series.

❑ **Note:** The above equation gives the **reciprocal** of the total capacitance.

For **only two** capacitors in series, then

$$C = \frac{C_1 C_2}{C_1 + C_2} \text{ farad}$$

Capacitors in parallel

$$C = C_1 + C_2 + \ldots + C_n \text{ farad}$$

Energy stored in a capacitor

$$W = \frac{1}{2}CV^2 \text{ joule}$$

$$\text{or } W = \frac{1}{2}QV \text{ joule}$$

Constants

Permittivity of free space, $\varepsilon_0 = 8.854 \times 10^{-12}$ farad/metre

Absolute permittivity of a dielectric, $\varepsilon = \varepsilon_0 \varepsilon_r$ farad/metre

Relative permittivity for various dielectrics

Material	ε_r (no units)
Air	1.0
Glass	7.5
Mica	6.2
Paraffin wax	2.2
Perspex	3.5
Polystyrene	2.55
PVC	4.5
Teflon	2.1

Assignment Questions _____

2.1 The capacitor in a camera flashgun attachment has a capacitance of 100 μF. When the flash is operated the capacitor becomes fully discharged, and it is then automatically recharged from two 1.5 V batteries which are connected in series. Calculate
(a) the energy stored in the fully charged capacitor, and
(b) the time taken to recharge between flashes if the average recharging current is 50 μA.

2.2 A pair of parallel plates having dimensions 50 mm × 40 mm are separated by a sheet of mica of thickness 0.8 mm. When an average charging current of 3 μA is applied for a period of 4 ms determine
(a) the p.d. developed between the plates
(b) the force exerted between the plates, stating whether this is a force of attraction or repulsion, and
(c) the potential gradient between the plates.

2.3 The station tuning capacitor in a radio set consists of an air-dielectric variable capacitor having 24 interleaved plates. Each pair of adjacent plates are 1 mm apart. At one extreme of rotation of the movable plates the effective area of overlap between fixed and movable plates is 40 cm^2. At the other extreme the overlap is 4 cm^2. Determine
(a) the maximum and minimum values of capacitance, and
(b) the maximum values of charge and energy stored when the capacitor is connected to a 9 V d.c. supply.

2.4 A switchable capacitor bank is shown in Fig. 2.1, where all capacitor values are in microfarads. For the following combinations of switch settings calculate the capacitance available between terminals X and Y

Fig. 2.1.

(a) switch A in position '1'; switch B open; switch C open
(b) switch A in position '1'; switch B closed; switch C open
(c) switch A in position '2'; switch B closed; switch C open
(d) switch A in position '2'; switch B closed; switch C closed
(e) switch A in position '1'; switch B open; switch C closed
(f) with the switches set as in (d) above, and with 24 V applied across terminals X and Y, calculate the p.d. across and energy stored by the 10 μF capacitor.

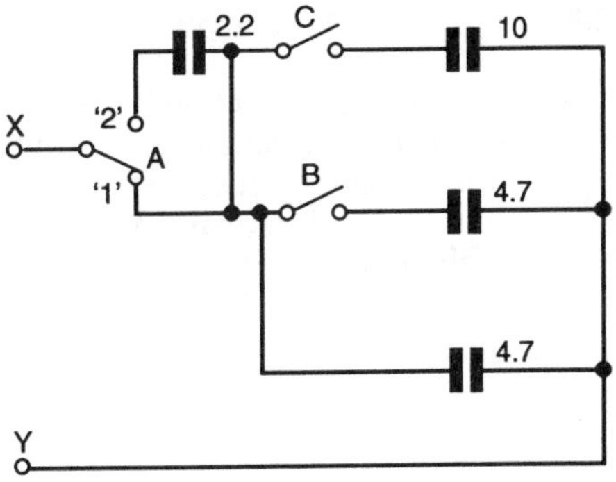

2.5 An insulating material has a dielectric strength of 100 kV/m. Three different samples of this material are to be tested by subjecting them to a p.d. of 250 V. If the three samples have thicknesses of (a) 1.5 mm, (b) 2.5 mm, and (c) 4 mm, respectively, determine the likely results of the tests.

2.6 A sheet of glass 2.5 mm thick separates two square aluminium plates which are connected to a 400 V d.c. source. Under this condition the electric flux produced in the glass is 21.52 nC. Calculate the following

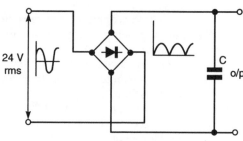

(a) the electric flux density in the glass
(b) the dimensions of the plates, and
(c) the capacitance of this arrangement.

2.7 A fullwave rectifier circuit operated from a 24 V root-mean-square (r.m.s.) a.c. supply is shown in Fig. 2.2. The smoothing capacitor, C, must be

Fig. 2.2.

capable of storing a peak energy of at least 8.5 J. Details of capacitors available are given below. Choose the capacitor that will be suitable for this purpose, and support your choice with the relevant calculations.

Capacitor No.	Value (μF)	Max. voltage (V)
1	22 000	30
2	10 000	63
3	15 000	40
4	10 000	25

2.8 An a.c. amplifier circuit employing a common emitter connected transistor is shown in Fig. 2.3.
(a) Explain the function the four resistors and three capacitors.
(b) The normal voltage gain of this circuit is 85, but when tested it was found to be only 2. This large reduction in gain is thought to be due to the failure of one of the seven passive components. Explain which one of these components is most likely to be at fault, and the manner of its failure.

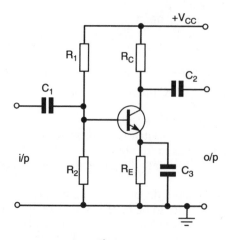

Fig. 2.3.

2.9 A simple a.c. amplifier circuit is illustrated in Fig. 2.4
   (a) sketch a typical frequency response curve for this amplifier, and
   (b) explain the reasons for the reduction in amplifier gain at both low and high frequencies.

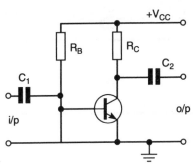

Fig. 2.4.

2.10 A capacitive linear displacement transducer utilises two concentric hollow aluminium cylinders as illustrated in Fig. 2.5. The inner cylinder has a diameter of 6 mm, whilst that of the outer cylinder is 7 mm, the dielectric being the air-gap between them. Each cylinder has a length of 3 cm. The displacement to be measured causes the inner cylinder to move axially within the outer cylinder. Maximum displacement causes the two cylinders to completely overlap. If the total range of displacement is 25 mm then calculate the maximum and minimum values of capacitance that result.

displacement

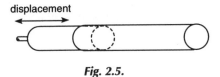

Fig. 2.5.

3 Magnetic Circuits and Electromagnetism

Equations

Magnetic flux density

$$B = \frac{\phi}{A} \text{ tesla (weber/metre}^2)$$

$$B = \mu_0 \mu_r H \text{ tesla}$$

Magnetomotive force (m.m.f.)

$$F = NI \text{ ampere turn}$$

Magnetic field strength

$$H = \frac{F}{l} = \frac{NI}{l} \text{ ampere turn/metre}$$

Reluctance

$$S = \frac{F}{\phi} \text{ ampere turn/weber}$$

$$S = \frac{l}{\mu_0 \mu_r A} \text{ ampere turn/weber}$$

For a magnetic circuit having 'n' sections in series

$$S = S_1 + S_2 + \ldots + S_n \text{ ampere turn/weber}$$

$$F = F_1 + F_2 + \ldots + F_n \text{ ampere turn}$$

Magnetic flux

$$\phi = \frac{F}{S} = \frac{NI}{S} \text{ weber}$$

Induced e.m.f.

$$e = -N\frac{d\phi}{dt} \text{ volt}$$

$$e = Blv \sin\theta \text{ volt}$$

Self-induced e.m.f.

$$e = -L\frac{di}{dt} = -N\frac{d\phi}{dt} \text{ volt}$$

Mutually-induced e.m.f.

$$e_2 = -M\frac{di_1}{dt} = -N_2\frac{d\phi_2}{dt} \text{ volt}$$

Self inductance

$$L = \frac{N^2}{S} = \frac{\mu_0\mu_r N^2 A}{l} \text{ henry}$$

Mutual inductance

$$M = \frac{N_2\phi_2}{I_1} \text{ henry}$$

For two coupled coils

$$M = k\sqrt{L_1 L_2} \text{ henry, where } 0 \leqslant k \leqslant 1$$

Total inductance of two coils in series

$$L = L_1 + L_2 \pm 2M \text{ henry}$$

Energy stored in a magnetic field

$$W = \tfrac{1}{2} LI^2 \text{ joule}$$

Energy stored by two coils in series

$$W = \tfrac{1}{2} L_1 I_1^2 + \tfrac{1}{2} L_2 I_2^2 \pm MI_1 I_2 \text{ joule}$$

Force on a current-carrying conductor

$$F = BIl \sin \theta \text{ newton}$$

Force between two current-carrying conductors

$$F = \frac{2 \times 10^{-7} I_1 I_2}{d} \text{ newton/metre}$$

Force between magnetised surfaces

$$F = \frac{B^2 A}{2\mu_0} \text{ newton}$$

Torque exerted on a current-carrying coil in a magnetic field

$$T = BANI \text{ newton metre}$$

Figure of merit for a voltmeter

$$\text{Figure of merit} = \frac{1}{I_{fsd}} \text{ ohm/volt}$$

Constants _____

Permeability of free space, $\mu_0 = 4\pi \times 10^{-7}$ henry/metre

Absolute permeability of a material, $\mu = \mu_0 \mu_r$ henry/metre

B/H data for certain materials are given in the form of graphs as in Fig. 3.1.

Assignment Questions _____

❑ **Note:** Where magnetic data are required, and are not given in a particular question, then refer to the B/H graphs of Fig. 3.1 opposite.

3.1 A permanent magnet having a rectangular cross-section of 10 mm × 15 mm produces a flux density of 1.35 T. This magnet is to be used to provide a flux density of 0.75 T between a pair of pole pieces as shown in Fig. 3.2 overleaf. If the air-gap between the pole pieces is 8 mm long, determine
(a) the cross-sectional area of the outer faces of the pole pieces
(b) the effective magnetic field strength in the air-gap, and
(c) the reluctance of the air-gap.

3.2 A single cable 20 m long is carrying a current of 52 A.
(a) Determine the magnetic field strength and flux density at a distance of 10 cm from its centre.

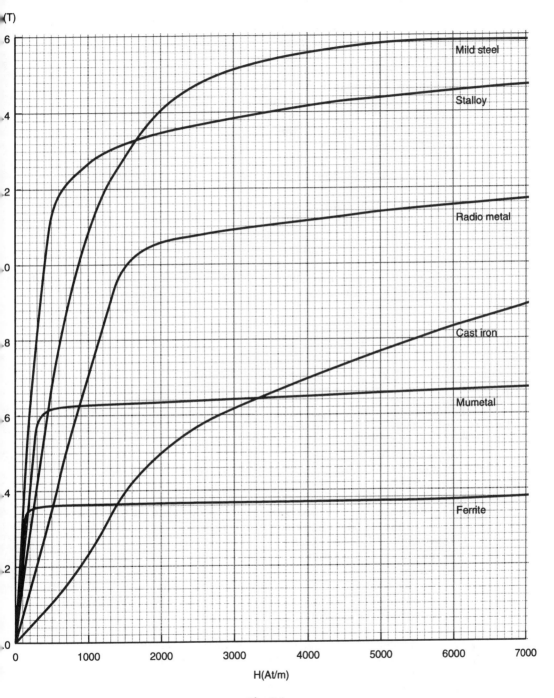

Fig. 3.1.

(b) A second identical cable, carrying the same value of current (52 A), but in the opposite direction, is situated such that the distance between the centres of the two cables is 1 cm. Calculate the force existing between the two cables, and state whether this is a force of attraction or repulsion.

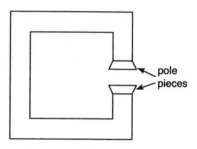

pole
pieces

Fig. 3.2.

3.3 An aircraft has a wingspan of 30 m. Given that the vertical component of the Earth's magnetic field is 42.6 μT, determine the e.m.f. induced in the wings when the aircraft is travelling at a speed of 820 km/h.

3.4 A magnetic circuit made from mumetal carries a flux density of 0.64 T. Determine
(a) the relative permeability under this condition, and
(b) the flux in the circuit if the cross-sectional area is 14 cm^2.

3.5 An iron toroid of length 40 cm is tested in order to determine its magnetic properties. This testing involved winding a 100-turn coil of wire around it, increasing the coil current in steps up to 5 A in one direction; reducing the current back to zero, and then reversing it. For each current setting the magnetic flux density was measured, the results being as shown in Table 3.1.

Table 3.1

I(A)	0.5	1.0	1.5	2.0	2.5	3.0	3.5	4.0	4.5	5.0	current
B(mT)	95	180	245	288	320	347	360	378	390	400	increasing
B(mT)	167	223	268	306	335	355	373	385	395	400	decreasing
B(mT)	10	−64	no further results obtained								reversed

Using these results
(a) determine and tabulate the corresponding values for the magnetic field strength (H)
(b) plot the sections of the B/H curve
(c) using **decreasing** current values, determine and tabulate the corresponding values for the toroid's relative permeability (μ_r)
(d) from the plotted graphs, determine
  (i) the residual flux density
  (ii) the coercive force, and
  (iii) the maximum value of coil current allowed if the relative permeability is to be $\geqslant$ 350. Note: Use values of B and H from the curve obtained with **decreasing** current values.

3.6 The core of a large inductor is constructed with a number of L-shaped stalloy laminations as shown in Fig. 3.3. The coil has 1500 turns and under normal use the flux density in the core is 0.4 T. For this condition calculate
(a) the flux in the core
(b) the coil current
(c) the relative permeability of the stalloy

(d) the coil inductance, and

(e) the back e.m.f. induced if the coil current is reduced to zero in a time of 4 ms.

3.7 The inductor specified in question 3.6 is accidentally dropped from a workbench, with the result that the two L-shaped sections are jarred slightly apart. The effective air-gap thus introduced into the circuit is 0.8 mm. If the core flux density is to maintained at 0.4 T, determine

(a) the value to which the coil current must be increased

(b) the total reluctance of the magnetic circuit, and

(c) the value of inductance now available.

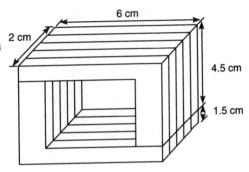

Fig. 3.3.

3.8 An inductor is constructed using 40 SWG constantan wire (diameter 0.122 mm) which is tight-wound onto a hollow plastic former of square cross-section as shown in Fig. 3.4. Determine

(a) the number of turns required to cover the former

(b) the length of wire required

(c) the coil resistance (refer to Constants in Chapter 1)

(d) the m.m.f. produced when a 24 V supply is connected across the coil

(e) the inductance value, and

(f) the inductance value if a tight-fitting ferrite core of effective length 120 mm and relative permeability 70 is included.

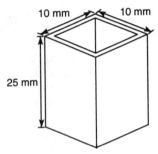

Fig. 3.4.

3.9 Figure 3.5 represents the cross-section of the magnetic circuit of a d.c. generator. The yoke and pole pieces of the machine are made from cast iron and the armature consists of mild steel laminations. The air-gap between each pole and the armature is 0.5 cm. The magnetising field winding is equally distributed between the two poles, each one carrying 500 turns. The effective axial length of the yoke and the armature is 6 cm. Determine the field current required to produce a flux of 1 mWb in the air-gaps. Note: you may take the mean fluxpath length across the armature to be 5.6 cm.

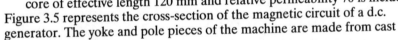

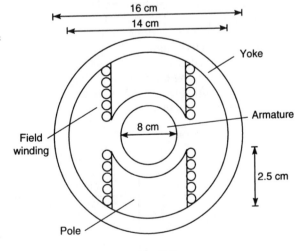

Fig. 3.5.

3.10 Figure 3.6 represents the physical
arrangement of an
electromagnetic relay. The relay
coil has 200 turns, and when
energised the magnetic field
produced attracts the hinged
spring-loaded armature, and
hence closes the contacts. The
air-gap between the core and the
armature, and that due to the
hinge arrangement is equivalent
to a total air-gap of 1.5 mm. The
mean length and cross-sectional
area of the magnetic circuit are
95 mm and 40 mm^2, respectively.
The force required to overcome
the spring is 2 N.

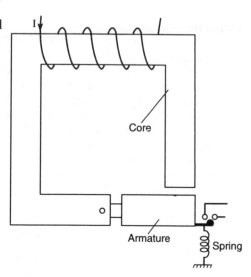

Core

Armature

Spring

Fig. 3.6.

(a) Using the *B/H* data given in
   Table 3.2, determine the
   minimum coil current
   required to operate the relay, and
(b) when the coil current is turned off it is found that the contacts do not open
   until the current has fallen to about 30% of its original value. Give a short
   explanation as to why this effect occurs.

Table 3.2

B (T)	0.1	0.2	0.3	0.4	0.5	0.6
H (At/m)	80	100	122	146	172	200

3.11 A series magnetic circuit consists of three sections
A, B and C, as shown in Fig. 3.7, where section C
has an air-gap in it. The coil wound on
to section A has 850 turns and is supplied with a
current of 1.2 A. The magnetic and physical
properties of the various sections are given in
Table 3.3. Determine

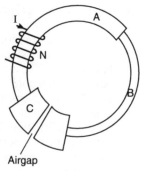

Airgap

Fig. 3.7.

(a) the flux density in each section
(b) the resulting inductance of the arrangement
(c) the inductance that would be obtained if the
   air-gap did not exist.

Table 3.3

Section	*l* (mm)	*A* (mm^2)	μ_r
A	80	120	400
B	60	40	250
C	50	200	600
Air-gap	1	40	1

3.12 The moving coil in a voltmeter is wound with 45 turns of 0.12 mm diameter copper wire onto an aluminium former of the dimensions shown in Fig. 3.8. The flux density in which the coil rotates is 0.2 T.

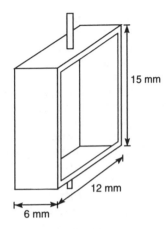

15 mm

12 mm

6 mm

Fig. 3.8.

(a) Determine

    (i) the restoring torque exerted by each of two contrawound spiral springs when the meter is carrying its full-scale deflection current of 150 μ A,

    (ii) the resistance of the coil, and

    (iii) the value of multiplier resistance required to enable the meter to give a full-scale deflection with 10 V applied to its terminals.

(b) The coil, on its former, rotates about a soft iron cylinder. Explain the purpose of this cylinder.

3.13 Two inductor coils are wound onto a ferrite toroid of circular cross-section having the dimensions shown in Fig. 3.9. Coil A has 300 turns whilst coil B has 150 turns. The coils may be used either individually or connected in series. Given that the coupling coefficient between the coils is 0.95, and that the relative permeability of the ferrite is 1800, determine

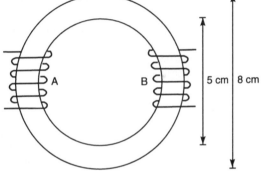

A   B   5 cm | 8 cm

Fig. 3.9.

(a) the inductance of each coil

(b) the possible values of inductance available when they are connected in series, and

(c) the series combination required in order to store an energy of 0.273 J when carrying a current of 2 A.

3.14 In a laboratory experiment, two coils are wound onto a ferromagnetic core such that 82% of the flux produced by the first coil links with the second. The current passed through the first coil is increased uniformly from zero to 2.5 A in a time of 5 ms; is held constant at 2.5 A for a further 6 ms, and is then reversed in a time of 8 ms. A dual beam oscilloscope is connected to monitor the e.m.f.s generated in the two coils.

(a) Draw, to scale, graphs to show the waveforms that you would expect to be displayed on the oscilloscope if the inductance of coil 1 is 120 mH and that of the second coil is 70 mH, and

(b) calculate the approximate number of turns on each of the coils if a current of 2.5 A through coil 1 produces a core flux of 121.36 μWb.

3.15 The no-volt relay (NVR) in a d.c. faceplate motor starter holds the control arm in the 'ON' position when carrying the motor field current of 4.5 A. The core and armature of the NVR are constructed from mild steel, and have the dimensions shown in Fig. 3.10. Due to the action of the control spring, the force required to hold the arm in the 'ON' position is 50 N. Determine the number of turns of wire needed on the core if the effective air-gap between the core and the armature is a total of 0.1 mm.

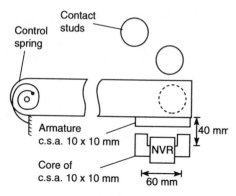

Fig. 3.10.

3.16 A 'U'-shaped lifting magnet of uniform cross-section has two pole faces, each of area 100 cm². This electromagnet is used to pick up flat mild steel plates, having a mass of 250 kg and cross-sectional area 40 cm². Due to surface irregularities there is an effective air-gap between the pole faces and the plate of 3 mm at each pole face. The mean magnetic path length through the magnet is 500 mm, whilst that for the plate is 200 cm. The arrangement is illustrated in Fig. 3.11. *B/H* data for the magnet material are given in Table 3.4, and that for the mild steel plate may be obtained from Fig. 3.1. Calculate the current required for the 250-turn coil, wound around the central limb of the electromagnet, in order to lift one plate. Note: To convert kg force to newton force multiply by 9.81.

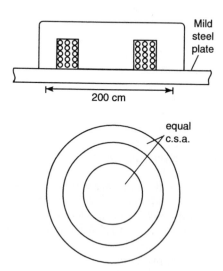

Fig. 3.11.

Table 3.4

B (T)	0.3	0.4	0.5	0.6	0.7
H (At/m)	400	600	850	1200	1600

3.17 Figure 3.12(a) shows the magnetic circuit of a relay. It is constructed from cast steel and the cross-section throughout is 1 cm × 1 cm. The relay is required to exert a force of 7 N in order to attract the armature. When the relay is de-energised the air-gap is restricted to 2.5 mm, and the pivot *P* can be assumed equivalent to an additional air-gap of 0.5 mm. The relay coil is to be of 0.327 mm diameter copper wire wound onto a hollow cylindrical former fitting over part A B. This former is shown in Fig. 3.12(b). The relay coil is to be supplied from a switching transistor that provides an output of 2.5 W at 16 V. Data for the cast steel is given in Table 3.5. Determine
(a) the m.m.f. required to operate the relay
(b) the number of turns of wire required, and
(c) the depth *d* of the coil.

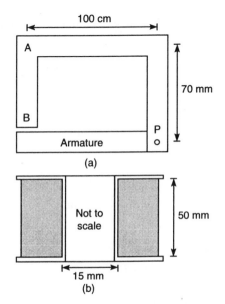

Fig. 3.12.

Table 3.5

B (T)	0.24	0.35	0.45	0.52
H (At/m)	100	200	300	400

4 Alternating Quantities and A.C. Circuits

Equations

Relationship between frequency and number of poles

$$f = np \text{ hertz}$$

where n = speed of rotation in rev/s and p = number of pole *pairs*.

Angular velocity and frequency

$$\omega = 2\pi f \text{ rad/s}$$

Period, or Periodic time

$$T = \frac{1}{f} \text{ second}$$

Standard expressions

$$v = V_m \sin(\omega t) \text{ volt}$$
$$v = V_m \sin(2\pi f t) \text{ volt}$$

angles in **radians**

$$v = V_m \sin\theta \text{ volt}, \text{ where } \theta \text{ is in degrees}$$

Series single-phase circuits (using circuit current as reference phasor)

Affect of circuit components

Pure resistance: $I = \dfrac{V}{R}$ amp; $\phi = 0°$; $P = I^2 R$ watt

Pure inductance: $I = \dfrac{V}{X_L}$ amp; $\phi = +90°$ or $+\pi/2$ rad; $X_L = 2\pi f L$ ohm; $P = $ zero

Pure capacitance: $I = \dfrac{V}{X_C}$ amp; $\phi = -90°$ or $-\pi/2$ rad; $X_C = \dfrac{1}{2\pi f C}$ ohm; $P = $ zero

Impedance

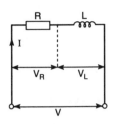

$$Z = \sqrt{R^2 + X_L^2} \text{ ohm}$$

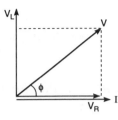

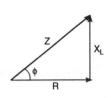

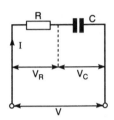

$$Z = \sqrt{R^2 + X_C^2} \text{ ohm}$$

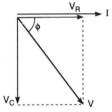

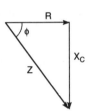

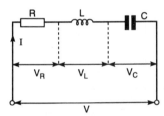

$$Z = \sqrt{R^2 + (X_L + X_C)^2} \text{ ohm}$$

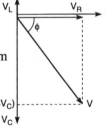

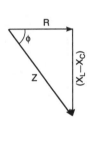

and in every case $I = V/Z$ amp

Power, power factor (p.f.) and power triangle

$$P = V_R I \text{ watt} = I^2 R \text{ watt} = VI \cos \phi \text{ watt}$$

$$\text{Power factor, } \cos \phi = \frac{\text{true power}}{\text{apparent power}} \text{ or } \frac{R}{Z}$$

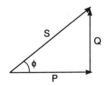 or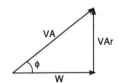

Series resonance

$$f_0 = \frac{1}{2\pi\sqrt{LC}} \text{ hertz}; \quad Z = R; \quad \phi = 0°$$

Voltage magnification or Q-factor, $Q = \dfrac{V_C}{V} = \dfrac{V_L}{V}$

Bandwidth, $B = \dfrac{f_0}{Q} = \dfrac{1}{R}\sqrt{\dfrac{L}{C}}$ hertz

Parallel single-phase circuits (using applied voltage V as reference phasor)

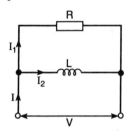

$I = \sqrt{I_1^2 + I_2^2}$ amp; $Z = \dfrac{V}{I}$ ohm

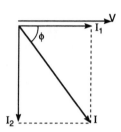

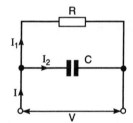

$I = \sqrt{I_1^2 + I_2^2}$ amp; $Z = \dfrac{V}{I}$ ohm

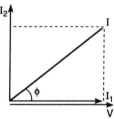

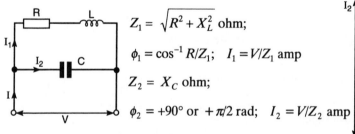

$Z_1 = \sqrt{R^2 + X_L^2}$ ohm;

$\phi_1 = \cos^{-1} R/Z_1$; $I_1 = V/Z_1$ amp

$Z_2 = X_C$ ohm;

$\phi_2 = +90°$ or $+\pi/2$ rad; $I_2 = V/Z_2$ amp

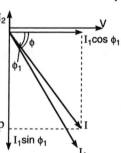

$I =$ **phasor** sum of I_1 and I_2; and $Z = V/I$ ohm

Parallel resonance

$$f_0 = \dfrac{1}{2\pi}\sqrt{\dfrac{1}{LC} - \dfrac{R^2}{L^2}}\ \text{hertz}$$

Circuit or dynamic impedance, $Z_D = \dfrac{L}{CR}$ ohm

Current magnification or Q-factor, $Q = \dfrac{I_C}{I} = \dfrac{I_L}{I}$ ohm

$$\text{or } Q = \frac{\omega_0 L}{R} = \frac{1}{\omega_0 CR}$$

Bandwidth, $B = \dfrac{f_0}{Q} = \dfrac{1}{R}\sqrt{\dfrac{L}{C}}$ hertz

Three-phase circuits

Line and phase quantities

$$V_L = V_{ph};\ I_L = \sqrt{3}\,I_{ph} \text{ in delta connection}$$
$$V_L = \sqrt{3}\,V_{ph};\ I_L = I_{ph} \text{ in star connection}$$

Power and power measurement

$$P = \sqrt{3}\,V_L I_L \cos\phi \text{ watt}$$

and for the two-wattmeter method of measurement

$$P = P_1 + P_2 \text{ watt}$$

where

$$P_1 = V_L I_L \cos(30° - \phi) \text{ watt}$$

and

$$P_2 = V_L I_L \cos(30° + \phi) \text{ watt}$$

The load phase angle, $\phi = \tan^{-1}\sqrt{3}\left(\dfrac{P_2 - P_1}{P_2 + P_1}\right)$ degree

Transistor h-parameters

Transistor input impedance, $h_{ie} \approx R_{in}$ ohm

Transistor output impedance, $h_{oe} \approx \dfrac{1}{R_{out}}$ siemen

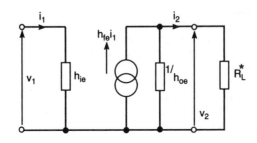

$$\text{Transistor current gain, } h_{fe} \approx \beta = h_{FE}$$

$$\text{Amplifier current gain, } A_i \approx \frac{h_{fe}}{1 + h_{oe}R_L^*} = \frac{i_2}{i_1}$$

where $R_L^* = $ the **effective a.c. load.**

$$\text{Amplifier voltage gain, } A_V \approx \frac{h_{fe}R_L^*}{h_{ie}(1 + h_{oe}R_L^*)} = \frac{v_2}{v_1}$$

Constants

Average and r.m.s. values

For a sine wave:
$$I_{av} = \frac{2}{\pi}I_m = 0.637I_m; \quad I = \frac{I_m}{\sqrt{2}} = 0.707I_m$$

For a square wave:
$$I_{av} = I = I_m$$

For a triangular wave:
$$I_{av} = \frac{I_m}{2} = 0.5I_m; \quad I = \frac{I_m}{\sqrt{3}} = 0.577I_m$$

Peak and form factors

For a sinewave: peak factor $= \sqrt{2} = 1.4142$; form factor $= 1.11$

For a squarewave: peak factor $=$ form factor $= 1.0$

For a triangular wave: peak factor $= \sqrt{3} = 1.7321$; form factor $= 1.155$

Assignment Questions

4.1 An alternating current was measured by a d.c. ammeter in conjunction with a full-wave rectifier. The ammeter reading was 15 mA. Assuming a sinusoidal waveform, calculate the r.m.s. and peak values for the current.

4.2 A voltage of 120 sin $100\pi t$ volt is applied across a circuit consisting of a diode in series with a 20 Ω resistor. Assuming the diode to be ideal, determine the peak and r.m.s. values of the resulting current.

4.3 A sinusoidal voltage displayed on an oscilloscope screen is illustrated in Fig. 4.1.
   (a) Determine the frequency, amplitude and r.m.s. value of this waveform if the timebase and Y-amplifier settings, respectively, were
       (i)   2 ms/cm; 5 V/cm
       (ii) 10 μs/cm; 0.5 V/cm
       (iii) 100 μs/cm; 0.1 V/cm.
   (b) Write down the standard expression for the voltage in each of the above cases. (You may assume a phase angle of zero.)

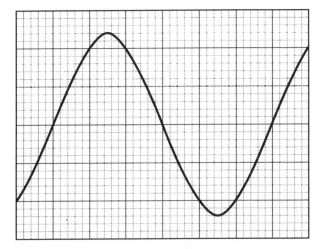

Fig. 4.1.

4.4 The p.d.s across two series components in an a.c. circuit are monitored on a dual beam oscilloscope. The waveforms displayed are illustrated in Fig. 4.2, and each channel is set to 50 ms/cm and 10 V/cm. Determine
(a) the r.m.s. value of each p.d.
(b) the phase angle between them, and
(c) the standard expression for the supply voltage.

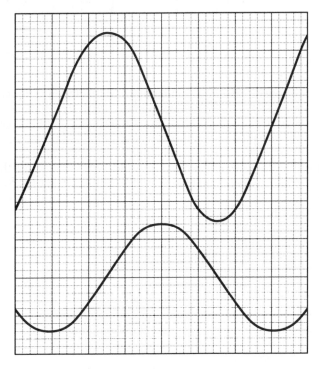

Fig. 4.2.

4.5   The branch currents in a parallel a.c. network are as follows. Calculate the total current supplied.

$$i_1 = 25 \sin 628t \text{ milliamp}$$

$$i_2 = 42 \sin (628t - \pi/4) \text{ milliamp}$$

$$i_3 = 60 \sin (628t + \pi/3) \text{ milliamp}$$

4.6   An inductor and a capacitor are subjected to the following tests.
(a) 24 V d.c. applied to the inductor produced a current of 0.48 A
(b) 24 V, 50 Hz applied to the inductor produced a current of 0.406 A
(c) 24 V, 1.5 kHz applied to the capacitor produced a current of 0.106 A.
Using these results determine the values for the two components.

4.7   As part of a practical test a student is presented with three sealed boxes, each one containing a single passive electrical component connected between a pair of external terminals. With the aid of a digital multimeter, a signal generator and a 12 V d.c. supply the student recorded the measurements shown in Table 4.1. Using these measurements determine the nature and value of each of the three boxed components.

Table 4.1

Applied voltage	Frequency (Hz)	Current flow (mA)		
		Box 1	Box 2	Box 3
12 V	d.c.	160.0	21.4	0
10 V	1 kHz	31.0	17.9	13.8
20 V	4 kHz	15.9	35.7	110.6

4.8   An a.c. circuit consists of a practical inductor connected in series with an ideal capacitor, across a 24 V, 200 Hz supply. The circuit current, measured with an ammeter, was found to be 0.268 A. The p.d.s across the inductor and capacitor were measured using a dual beam oscilloscope, and had peak-to-peak values of 54.5 V and 120.6 V, respectively. Using the timebase on the oscilloscope it was determined that the voltage across the inductor leads that across the capacitor by an angle of 164°. Using this data determine the values for the inductor and capacitor.

4.9   For the circuit shown in Fig. 4.3 calculate
(a) the value of the resistor R
(b) the resistance and inductance (r and L) of the coil, and
(c) the circuit power factor and the power dissipated.

4.10 Four 110 V, 60 W incandescent lamps (purely resistive) are connected in parallel. Determine the value of a series connected capacitor that will enable these lamps to operate at the

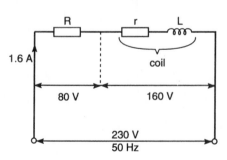

Fig. 4.3.

correct p.d. of 110 V when the complete circuit is connected to a 230 V, 50 Hz supply.

4.11 An a.c. motor having a resistance of 20 Ω and inductance of 0.05 H is connected in parallel with a 47 μF capacitor across a 230 V, 50 Hz supply. Calculate
  (a) the current drawn from the supply
  (b) the circuit power factor, and
  (c) the power dissipated.

4.12 For the circuit shown in Fig. 4.4, calculate
  (a) each branch current
  (b) the current drawn from the supply
  (c) the total power dissipation.

4.13 A resistor and capacitor are connected in series across a 50 V variable frequency a.c. supply. When the frequency is 40 Hz the circuit current is 117.6 mA, and at a frequency of 50 Hz the current is 142.1 mA. Determine the values for the resistor and capacitor.

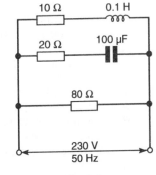

Fig. 4.4.

4.14 An e.m.f. of 280 sin (314*t* + π/4) volt is applied to an inductive circuit and the resulting current is 5.6 sin (314*t* − π/6) amp. Determine
  (a) the supply frequency
  (b) the resistance and inductance of the circuit, and
  (c) the power dissipated.

4.15 A series circuit is shown in Fig. 4.5. For this circuit, calculate
  (a) the resonant frequency, Q-factor and bandwidth
  (b) the p.d. across each of the three components when the circuit is connected to a 30 V supply at the resonant frequency.

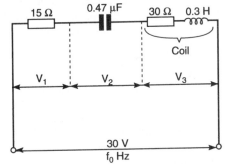

Fig. 4.5.

4.16 Figure 4.6 shows a series acceptor circuit having selectivity switchable between two preset levels. The circuit is required to have a resonant frequency of 5 kHz, with a minimum bandwidth of 100 Hz (i.e. with switch S in the closed position). Determine
  (a) the Q-factor with switch S closed
  (b) the required value of capacitor C, and
  (c) the value to which resistor *R* must be set in order to increase the bandwidth to 250 Hz.

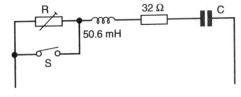

Fig. 4.6.

4.17 A coil of resistance 20 Ω and inductance 380 mH is connected in parallel with a circuit consisting of a 10 Ω resistor in series with a 22 μF capacitor. The complete circuit is supplied at 230 V, 50 Hz, and is shown in Fig. 4.7. Determine

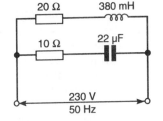

Fig. 4.7.

(a) the current flowing in each branch
(b) the total current drawn
(c) the power dissipated, and
(d) the value to which C must be changed in order for the circuit to have unity power factor.

4.18 A circuit consisting of a coil and a capacitor connected in series was tested by connecting the circuit across a variable frequency signal generator. The generator output was maintained constant at 500 mV, the frequency varied and the amplitude of the resulting current monitored. It was found that at a frequency of 7.5 kHz the current had reached its maximum value of 10 mA, and the p.d. across the capacitor was 25 V. From these results determine
(a) the circuit Q-factor
(b) the capacitor value
(c) the values for the resistance and inductance of the coil, and
(d) the minimum value of working voltage suitable for the capacitor.

4.19 A coil of resistance 12 Ω and inductance 0.15 H is connected in parallel with a 47 μF capacitor across a 10 V variable frequency supply. Calculate
(a) the frequency at which the circuit will behave as a non-reactive resistor
(b) the value of this circuit resistance, and
(c) the circuit current flow under this condition.

4.20 An inductor of resistance 400 Ω and inductance 318 μH is to be used as one of the two components in a parallel tuned circuit, the other being a capacitor. Determine
(a) the value of capacitor needed to achieve a tuned frequency of 1.5 MHz, and
(b) the circuit impedance at this frequency.

4.21 A tuned circuit bandpass filter circuit is illustrated in Fig. 4.8. In this circuit inductor L_1 has an inductance of 750 μH and resistance 5 Ω, and capacitor C_1 has a value of 338 nF. Inductors L_2 and L_3 each have values equal to one half of those for L_1, whilst capacitors C_2 and C_3 each have a value equal to twice that of C_1. For this circuit determine
(a) the centre pass frequency
(b) the output voltage V_0, at the centre frequency, when the input voltage V_i is 10 V, and

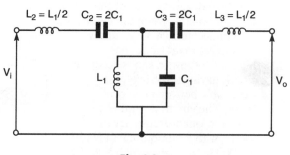

Fig. 4.8.

(c) the output voltage when the input is 10 V at a frequency of 5 kHz.

4.22 An office has the following loads connected to the 230 V, 50 Hz supply:
(i)   incandescent lamps drawing a current of 8.5 A at unity power factor
(ii) fluorescent lamps drawing a current of 6 A at power factor 0.8 leading
(iii) office machinery drawing a current of 9 A at power factor 0.7 lagging.
Calculate the total current, power factor, power and reactive voltampere taken from the supply.

4.23 For the loads specified in question 4.22, determine the value of power factor improvement capacitor needed to improve the overall power factor to unity.

4.24 A single-phase motor draws a current of 8.5 A at a power factor of 0.75 lagging from a 240 V, 50 Hz supply. Two identical capacitors connected in parallel form a capacitor bank, and this bank is connected across the motor terminals in order to improve the overall power factor. Determine the value of each capacitor in order to raise the p.f. to (a) unity, and (b) 0.9 lagging, and (c) explain the practical reasons why alternative (b) above is the better of the two options.

4.25 An a.c. supply provides the following loads:
   (i)   a heating load (resistive) of 18 kW
   (ii) a motor load of 50 kVA at a power factor of 0.6 lagging, and
   (iii) a further mixed load of 20 kW at a lagging power factor 0.8.
Calculate (a) the total load supplied, in kW and kVA, and (b) the kVAr rating of a power factor correction capacitor required to raise the overall power factor to 0.92 lagging.

4.26 A 20 kVA single-phase motor has a power factor of 0.8 lagging. A 5 kVA capacitor is connected in parallel with the motor. Determine the total kVA input and power factor when the motor is operating at full-load.

4.27 The transistor used in the amplifier circuit of Fig. 4.9 has the following small signal *h*-parameters:

$$h_{ie} = 2.5 \text{ k}\Omega; h_{fe} = 110; h_{oe} = 25 \text{ }\mu\text{S}$$

Sketch the complete *h*-parameter equivalent circuit and hence (or otherwise) determine the value of input signal generator e.m.f. *E* required so that a power of 2 mW is dissipated in the 15 kΩ load resistor. The reactances of the coupling capacitors may be ignored.

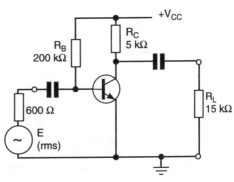

Fig. 4.9.

4.28 The output from a wire resistance strain gauge bridge circuit is fed to the input of a single-stage *RC* transistor amplifier. The bridge supply, *E* is 5 V r.m.s. and you may assume that the input resistance of the amplifier circuit has negligible loading effect on the bridge circuit. You may also assume that the reactances of the capacitors are negligible at the signal frequency. Under operating conditions the resistances of the strain gauges are as shown in Fig. 4.10.

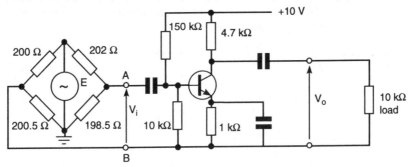

Fig. 4.10.

(a) Determine the p.d. between terminals A and B
(b) Sketch the *h*-parameter equivalent circuit for the amplifier, and hence (or otherwise) calculate the power delivered to the 10 kΩ load resistor. The small signal *h*-parameters for the transistor are:

$$h_{ie} = 2 \text{ k}\Omega; \ h_{fe} = 60; \ h_{oe} = 20 \ \mu\text{S}$$

4.29 A preamplifier which is used to boost a 20 μA r.m.s. signal from a current source is shown in Fig. 4.11. All resistors used have a tolerance of ±10%, and the transistor's small signal *h*-parameters are within the ranges given below:

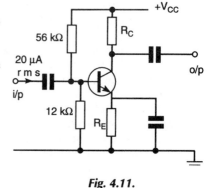

	Min.	Max.
h_{ie}	2.1 kΩ	4.4 kΩ
h_{fe}	75	220
h_{oe}	may be considered to be negligible	

(a) Sketch the *h*-parameter equivalent circuit and hence (or otherwise) determine the value of collector load resistor R_C that will result in an output voltage which will never be less than 6 V r.m.s. for any combination of transistor or resistor selected. Specify which of the following listed resistance values you would choose in practice:

Fig. 4.11.

5.6 kΩ; 6.2 kΩ; 6.8 kΩ; 7.5 kΩ

(b) Using this value determine the maximum possible output power which the amplifier can deliver to a load which is matched to the nominal value of the resistor calculated in part (a) above.

4.30 The stator of a three-phase induction motor contains three identical windings, each of resistance 25 Ω and inductance 0.125 H. These stator windings are connected to a 400 V, 50 Hz, three-phase supply via a star/delta starter arrangement. Calculate the stator line and phase currents, and the power drawn from the supply, when the windings are connected (a) in star, and (b) in delta configuration.

4.31 A 15:1 step-down, delta/star three-phase transformer supplies a star-connected load. Each phase of the load has an impedance of 16 Ω and a power factor of 0.65 lagging. The primary of the transformer is connected to a 6 kV, 50 Hz, three-phase supply, and the complete circuit arrangement is illustrated in Fig. 4.12. Determine

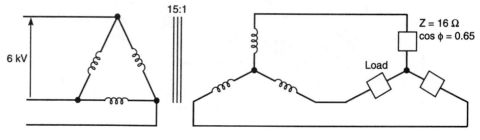

Fig. 4.12.

(a) the load phase and line currents

(b) the phase and line currents for the transformer primary, and

(c) the power drawn from the supply.

4.32 Three coils are connected in delta configuration to a three-phase, 400 V, 50 Hz supply, and draw a line current of 4.5 A at a lagging power factor of 0.8.

(a) Calculate the resistance and inductance of each coil, and

(b) if the coils are now star-connected to the same supply, calculate the line current drawn and the power dissipated.

4.33 A three-phase 50 Hz ideal transformer is used to supply a load of 400 kW at a lagging power factor of 0.8. The transformer secondary windings are delta-connected and has 400 turns/phase. The primary windings have 1525 turns/phase and are connected in star configuration. When the primary is connected to a 3.3 kV, 50 Hz, three-phase supply, calculate

(a) the voltages developed in both the primary and secondary windings, and

(b) the phase and line currents for the transformer and the load.

4.34 A ventilation fan is driven by a three-phase, delta-connected induction motor. The input power to the motor is measured using the two-wattmeter method, the two wattmeter readings being 450 W and 1050 W respectively. The supply is 400 V at 50 Hz. Determine

(a) the input power to the motor

(b) the motor power factor, and

(c) the motor line and phase currents.

4.35 A three-phase, delta-connected synchronous motor develops an output of 15 kW when operated from a 400 V, 50 Hz supply. If the machine has an efficiency of 87% and a power factor of 0.8, calculate

(a) the line current drawn

(b) the motor phase current, and

(c) the readings that would be indicated by two wattmeters connected to measure the total power input.

4.36 A three-phase, four-wire unbalanced load is illustrated in Fig. 4.13. The loads connected to the red, yellow and blue lines are 12 kW, 10 kW and 7 kW, respectively, each load being non-reactive. Determine

(a) the current in each line, and

(b) the current in the neutral conductor.

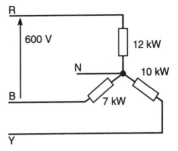

Fig. 4.13.

4.37 A factory has the following three-phase, four-wire loads, with a lagging power factor of 0.85 in each phase: red phase 45 A; yellow phase 55 A; blue phase 65 A. Given that the supply voltage is 415 V, determine

(a) the current in the neutral conductor, and

(b) the total power dissipated.

4.38 A three-phase, 400 V system has the following loads connected in delta
configuration: between red and yellow lines a resistive load of 125 Ω; between
yellow and blue lines an inductive reactance of 50 Ω, with negligible resistance;
between blue and red lines a capacitive reactance of 40 Ω. The circuit
arrangement is shown in Fig. 4.14. Calculate
(a) the phase currents, and
(b) the line currents.

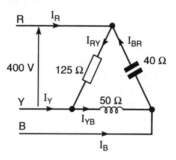

Fig. 4.14.

4.39 Three coils, each of resistance 25 Ω and reactance 15 Ω are connected in delta
to a three-phase, 400 V, 50 Hz supply.
(a) Determine
   (i) the line current drawn
   (ii) the power supplied, and
   (iii) the load power factor
(b) Three identical capacitors are connected in delta across this supply so as to
correct the load power factor to 0.95 lagging. Determine
   (i) the value of each capacitor, and
   (ii) the line current drawn from the supply under these conditions.

5 Networks, D.C. Transients and Feedback Amplifiers

Equations and Theorems

Constant voltage and constant current sources

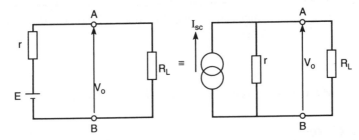

Where I_{sc} is the current that would flow between the *short-circuited* terminals A and B of the voltage source.

Thévénin's theorem

Where R_0 is the network impedance measured between terminals A and B of the network, with all sources replaced by their internal impedances. E_o is the **open-circuit** voltage that would be developed between terminals A and B

Norton's theorem

Where R_0 is the network impedance as defined for Thévénin's theorem, and I_{sc} is the current that would flow between the *short-circuited* terminals A and B of the network.

Maximum power transfer theorem

Maximum power is transferred to a load when the load impedance is equal to the internal impedance of the source, as illustrated below.

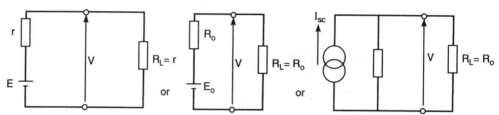

When a matching transformer is used to obtain maximum power transfer, then

$$R_p = \left(\frac{N_p}{N_s}\right)^2 R_L$$

where R_p is the secondary resistance referred back to the primary.

Decibel notation

Power gain, $A_p = 10\log\dfrac{P_o}{P_i}$ decibel

Voltage gain, $A_V = 20\log\dfrac{V_o}{V_i}$ decibel

Current gain, $A_i = 20\log\dfrac{I_o}{I_i}$ decibel

provided that the input and output resistance are the same value.

A *power* of 1 dB $= 1\,\text{mW}$

C-R series circuit

Time constant, $\tau = CR$ second

For charging:

Initial current, $I_0 = \dfrac{E}{R}$ amp

$i = I_0 e^{-t/\tau}$ amp

$V_C = E(1 - e^{-t/\tau})$ volt

$V_R = E - V_C = E e^{-t/\tau}$ volt

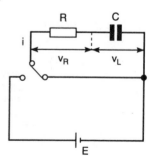

For discharge:

$$i = -I_0 e^{-t/\tau} \text{ amp}$$

$$v_C = v_R = E e^{-t/\tau} \text{ volt}$$

L-R series circuit:

Time constant, $\tau = \dfrac{L}{R}$ second

For increasing current:

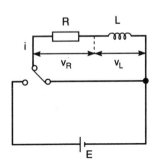

Final steady current, $I = \dfrac{E}{R}$ amp

$$i = I(1 - e^{-t/\tau}) \text{ amp}$$

$$v_L = E\, e^{-t/\tau} \text{ volt}$$

$$v_R = E(1 - e^{-t/\tau}) \text{ volt}$$

For decreasing current:

$$i = I\, e^{-t/\tau} \text{ amp}$$

$$v_R = v_L = E e^{-t/\tau} \text{ volt}$$

Differentiating and integrating networks

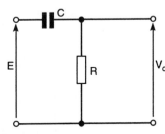

$V_o \approx CR\dfrac{\mathrm{d}E}{\mathrm{d}t}$ volt, provided that $\tau \leqslant T/20$,

            i.e. a relatively short time
constant.

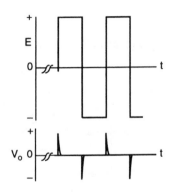

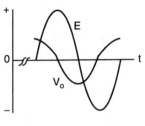

$$V_o \approx \frac{1}{CR} \int E dt \text{ volt, provided that } \tau \geqslant 5T,$$

i.e. a relatively long time constant.

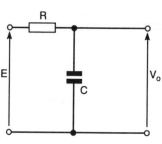

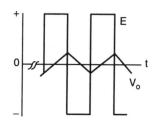

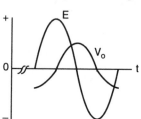

Differentiating and integrating operational amplifiers

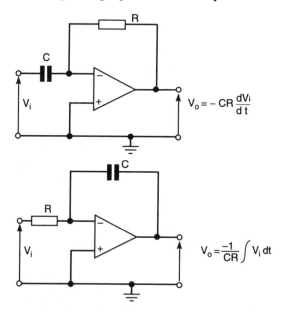

$$V_o = - CR \frac{dV_i}{dt}$$

$$V_o = \frac{-1}{CR} \int V_i \, dt$$

❑ **Note:** The above amplifier circuits each utilise the inverting (−) input, hence the minus sign associated with the output expression.

Negative feedback (nfb) amplifiers

Voltage gain with negative feedback,

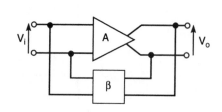

$$A'_v = \frac{A_v}{1 + \beta A_v}$$

where

$$A_v = \text{gain without feedback, and}$$

$$\beta = \text{feedback fraction.}$$

$$\text{or}, A'_v = 20 \log A_v - 20 \log(1 + \beta A_v) \text{ dB}$$

Cut-off (−3 dB) frequencies with nfb: $f'_1 = f_1/(1 + \beta A_v)$ hertz
$f'_2 = f_2(1 + \beta A_v)$ hertz

where f_1 and f_2 are those without nfb

Input and output resistances with nfb:

nfb applied in parallel with the input: $R'_{in} = R_{in}/(1 + \beta A_v)$ ohm

nfb applied in series with the input: $R'_{in} = R_{in}(1 + \beta A_v)$ ohm

where R_{in} is without nfb

nfb derived in parallel with the output: $R'_o = R_o/(1 + \beta A_v)$ ohm

nfb derived in series with the output: $R'_o = R_o(1 + \beta A_v)$ ohm

where R_o is without nfb

Astable multivibrators

555 timer version

Period of output waveform,
$T = t_1 + t_2$, where

$$t_1 = 0.69 C_1 (R_1 + R_2)$$
second

and $t_2 = 0.69 C_1 R_2$ second

❑ **Note:** If $R_2 \gg R_1$ then $t_1 \approx t_2$ giving a 1:1 mark to space ratio.

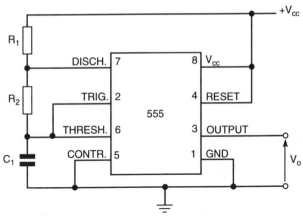

p.d. across C_1

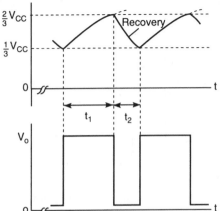

Op. Amp. version

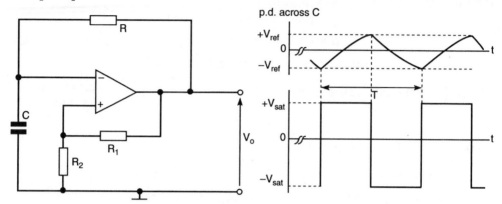

Period of output waveform, $T = t_1 + t_2$, where $t_1 = t_2$ second

$$T = 2CR\ln\frac{(1+\beta)}{(1-\beta)} \text{ second}$$

where
$$\beta = \text{feedback fraction} = \frac{R_2}{R_1 + R_2}$$

so
$$T = 2CR\ln\frac{(2R_2 + R_1)}{R_1} \text{ second}$$

and
$$T \approx 2CR \text{ second when } R_1 = 1.165\, R_2$$

Monostable multivibrators

555 timer version

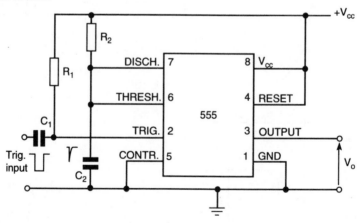

C_1 and R_1 differentiate input pulse to provide a negative-going spike of amplitude $\geq V_{cc}/3$ volt.

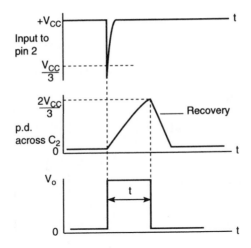

Output pulse duration, $t = 1.1\, C_2 R_2$ second.

Op. Amp. version

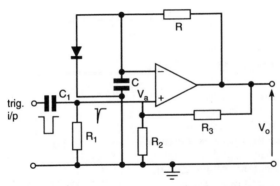

C_1 and R_1 differentiate the input to provide a negative-going spike of amplitude $\geqslant 0.6$ V.

Pulse time, $t = CR \ln \dfrac{1}{(1-\beta)}$ second

where

$$\beta = \text{feedback fraction} = \frac{R_2}{R_2 + R_3}$$

and if $R_2 = 1.717 R_3$, then $t = CR$ second

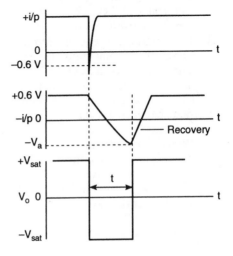

Schmitt trigger

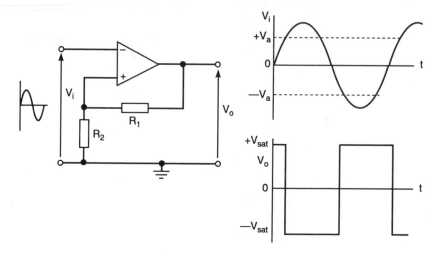

$$V_a = V_0 \frac{R_1}{R_1 + R_2} \text{ volt}$$

Assignment Questions

5.1 The output circuit of a bipolar common emitter connected transistor is represented as a constant current source as shown in Fig. 5.1. Determine and sketch the equivalent constant voltage circuit.

5.2 The output circuit of a FET is represented as a constant voltage source as shown in Fig. 5.2. Determine and sketch the equivalent constant current circuit.

5.3 The *h*-parameter equivalent for the input circuit of a bipolar transistor common emitter amplifier is shown in Fig. 5.3. According to the data sheet supplied, the transistor employed has an input impedance (h_{ie}) that may be any value between 1.8 kΩ and 3 kΩ. Using either Norton's or Thévénin's theorem, determine the value of base-emitter voltage (V_1) available for h_{ie} values at the two extremes and at the typical value of 2.4 kΩ.

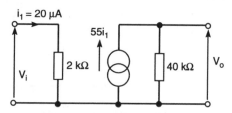

$i_1 = 20$ μA

Fig. 5.1.

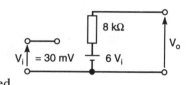

$V_i = 30$ mV   8 kΩ 6 V_i V_0

Fig. 5.2.

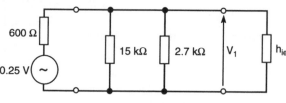

600 Ω 0.25 V   15 kΩ 2.7 kΩ V_1 h_{ie}

Fig. 5.3.

5.4 The bridge circuit shown in Fig. 5.4 can have a
number of different value resistive loads
connected between terminals A and B.
(a) Using either Thévénin's or Norton's theorem
determine the value of current through and
p.d. across these terminals when the load R_L
has a value of
(i) 5.5 Ω, (ii) 10 Ω, (iii) 15 Ω, and (iv) 20 Ω
(b) which one of the above load resistors will
result in the most power being transferred to
the load, and calculate the value of this
power
(c) check your answers to (a)(i) above by
applying Kirchhoff's laws, and
(d) determine the load current flowing if the 50 Ω
resistor is replaced by one of 60 Ω.

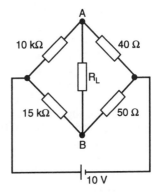

Fig. 5.4.

5.5 The power available at the receiving end of a transmission cable is only 6% of
that supplied at the sending end. Determine the cable power loss measured in
decibels.

5.6 An amplifier having a voltage gain of 27 dB has identical values of input and
output resistances. Calculate the input voltage that will result in an output
voltage of 4 V.

5.7 A block diagram of a transmission system is shown in Fig. 5.5, where each
block indicates the value of gain or loss in decibels. Determine
(a) the overall voltage gain or loss, and
(b) the output voltage if 15 mV is applied at the input.

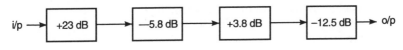

Fig. 5.5.

5.8 The signal level at a tape recorder head output is 850 μW. Calculate the output
noise power if the signal/noise ratio is 25 dB.

5.9 An amplifier having a power gain of 20 dB is supplied with an input power of
0.25 mW. Calculate the amplifier signal/noise ratio if this input results in an
output noise power of 0.3 μW.

5.10 A power of 1 μW is applied to the input of the system shown in Fig. 5.6.
Determine the power available at points A, B, and C, in (a) watts, and (b) dBm.

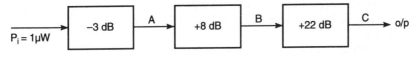

Fig. 5.6.

5.11 An amplifier has a power gain of 15 dB. When a power of 8 mW is applied to
its input, determine the output power in (a) watts, and (b) dBm.

5.12 An item of electronic equipment has both input and output resistances of
600 Ω. It provides an output current of 12 mA when the input voltage is 1.4 V.
Calculate the equipment voltage gain or loss, measured in decibels.

5.13 An amplifier has a power gain of 60 dB, and an input resistance of 75 Ω. The output resistance (including the load) is 150 Ω. Determine the p.d. across the load when an input of 10 mV is applied.

5.14 An amplifier has a voltage gain factor of 64, its input and output impedances are non-reactive and have values of 200 Ω and 600 Ω respectively. A transmission line connected to the amplifier output has a power loss of −10 dB. Express the amplifier power gain in dB and hence determine the overall power gain of the complete system, measured in decibels.

5.15 A 24 V d.c. supply may be connected, via a 4.7 kΩ resistor to either a 47 nF capacitor or a 0.5 H inductor by the use of a two-position switch as shown in Fig. 5.7. In each case, determine
(a) the time constant
(b) the current flowing 0.4 ms after the connection of the supply, and
(c) the stored energy at this time.
You may assume that the capacitor is initially uncharged.

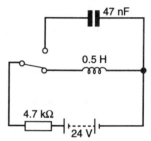

Fig. 5.7.

5.16 A 2.2 μF capacitor is connected in series with a 33 kΩ resistor. When this circuit is connected to a 50 V d.c. supply, calculate
(a) the initial rate of change of current
(b) the capacitor p.d. after 50 ms
(c) the time taken for the charging current to fall to 0.4 mA, and
(d) the approximate time taken for the capacitor to become fully charged.

5.17 A relay coil of inductance 200 mH and resistance 50 Ω is connected to a d.c. supply via a two-position switch as shown in Fig. 5.8. Assuming that the switch is initially in the position shown, determine
(a) the circuit time constant
(b) the current flowing 3 ms after the switch is operated
(c) the energy stored 3 ms after the switch is operated
(d) the time taken for the current to reach its final steady value, and
(e) the current flowing 2 ms after the switch is returned to its original position.

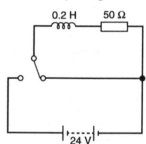

Fig. 5.8.

5.18 A relay coil of inductance 250 mH and resistance 50 Ω is operated by connecting it to a 24 V d.c. supply via a 50 Ω resistor and a switch, as shown in Fig. 5.9.

Initially the switch is in position '1' and the current through the relay coil is zero. After moving the switch to position '2' the relay armature is pulled in when the coil current reaches 165 mA.

Subsequent to steady-state conditions being attained, the switch is returned to position '1', and 5.5 ms later the relay armature is released.

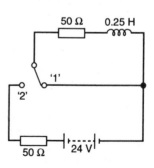

Fig. 5.9.

Determine
(a) the current drawn by the relay coil under steady-state conditions
(b) the time taken for the relay contacts to close after the supply is connected, and
(c) the coil current flowing at the instant that the relay contacts open.

5.19 For the network shown in Fig. 5.10, sketch to scale, on the same axes, the input and output waveforms, after several cycles have elapsed. Indicate all principal values on both the voltage and time axes.

To what value would the capacitor C have to be changed in order to achieve 'good' integration of the input waveform?

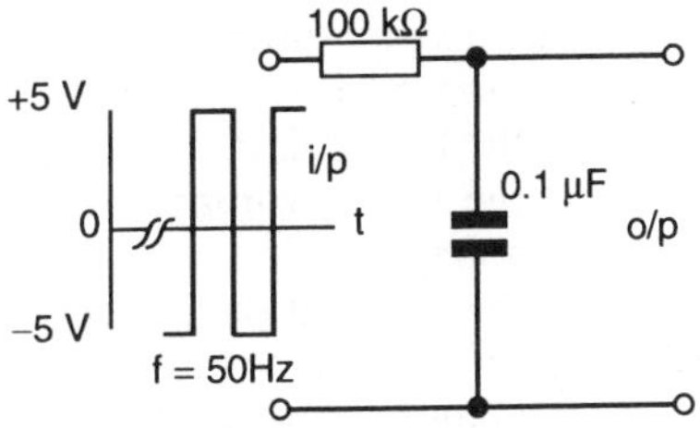

5.20 A symmetrical squarewave of amplitude ± 9 V and frequency 10 Hz is applied to the input of the integrating operational amplifier of Fig. 5.11. Sketch the input and output waveforms to scale, on the same axes indicating all principal values on both axes.

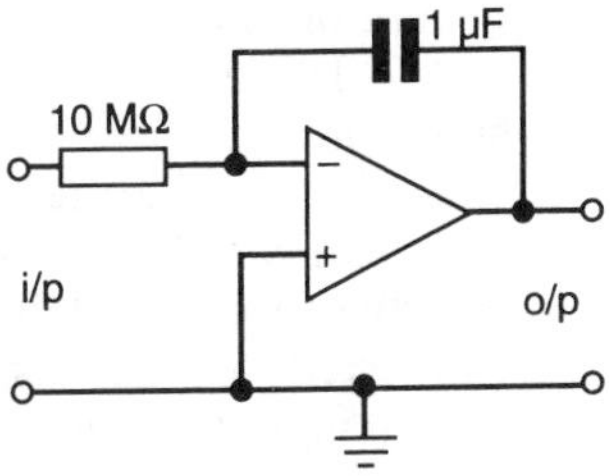

5.21 An integrating network consists of a 2 MΩ resistor connected in series with a 4.7 μF capacitor. A step-input voltage of 20 V is applied to the circuit and the output obtained from across the capacitor. Determine values for the ideal output, the actual output, and the percentage error introduced (a) 0.5 s, (b) 1.0 s, and (c) 4 s after the input is applied.

5.22 An integrating operational amplifier uses a feedback capacitor of 0.5 μF and an input resistor of 2.5 MΩ. The amplifier saturates (reaches its maximum possible output) at 15 V. If a constant input of -2.5 V is applied, determine the time taken for saturation to occur.

5.23 A alternating voltage of 3 sin 10t volts is applied to the input of an integrating operational amplifier. The passive components have values of $C = 1$ μF and $R = 0.5$ MΩ. Determine the expression for the alternating output waveform.

5.24 A simple filter and smoothing circuit consists of a 20 H inductor of resistance 100 Ω connected in series with a 50 μF capacitor as illustrated in Fig. 5.12. The input to this circuit is 120 + 15 sin 500t volts (i.e. a sinewave of amplitude 15 V superimposed on a d.c. level of 120 V). Determine
(a) the amplitude of the a.c. ripple present at the output, and
(b) the percentage of ripple to d.c. output.

5.25 A T-pad attenuator is shown in Fig. 5.13. For this circuit determine

Fig. 5.10.

Fig. 5.11.

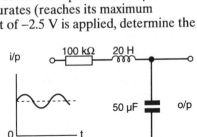

Fig. 5.12.

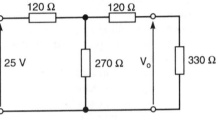

Fig. 5.13.

(a) the output voltage, V_o, and
(b) the voltage attenuation, in
   decibels

5.26 A source of e.m.f. 15 V and
   internal resistance 200 Ω is
   connected to a 750 Ω load via
   a resistive attenuator network
   as in Fig. 5.14. Determine
   (a) the load p.d, V_o
   (b) the attenuator input
      voltage, and
   (c) the attenuation, in decibels.

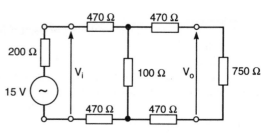

Fig. 5.14.

5.27 A 500 Ω load is to be supplied via an
   attenuator circuit such that the
   attenuation is –8 dB, and the
   network matches the load to an input
   generator of internal resistance
   600 Ω. The circuit arrangement is
   shown in Fig. 5.15. Determine the
   values required for resistors R_1 and R_2.

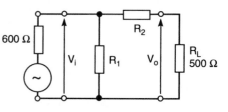

Fig. 5.15.

5.28 The attenuator pad of Fig. 5.16 has an
   effective input resistance of R ohm when
   connected to a load also of value R ohm.
   For an input of 40 V, calculate
   (a) the value of R
   (b) the p.d. across the load, and
   (c) the attenuation in decibels.

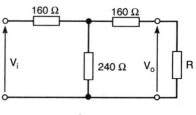

Fig. 5.16.

5.29 A negative feedback voltage amplifier
   has an open-loop gain factor of 100. If
   3% of its output is fed back to the input,
   determine
   (a) the gain with feedback (the closed-loop gain), and
   (b) the percentage change in closed-loop gain if, due to the ageing of
      components, the open-loop gain decreases by 10%.

5.30 An operational amplifier having an open-loop voltage gain factor of 10^5 is to be
   utilised as a negative feedback voltage amplifier having a closed-loop gain of
   80. Determine the feedback factor required.

5.31 The gain of a voltage amplifier without feedback is 50 dB. What will be the
   gain achieved if the negative feedback fraction is 0.6?

5.32 The negative feedback for an
   amplifier is derived using a potential
   divider across the output terminals as
   shown in Fig. 5.17. Given that the
   open-loop gain is 10^4, determine
   (a) the nominal gain with feedback
      applied, and
   (b) the minimum and maximum
      values for the closed-loop gain if
      the resistors used in the potential
      divider circuit have a tolerance of
      $\pm5\%$.

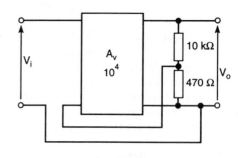

Fig. 5.17.

5.33 An amplifier has an open-loop gain 2×10^5, an input resistance of 0.2 MΩ and an output resistance of 75 Ω. When negative feedback is derived in parallel with its output and fed back in series with its input, the closed-loop gain becomes 500. Determine
(a) the feedback fraction, and
(b) the modified values for input and output resistance.

5.34 A negative feedback voltage amplifier has an open-loop gain of 80 dB and a bandwidth from 0 to 100 Hz. If the closed-loop gain is required to be 25 dB then determine
(a) the necessary value for the feedback fraction, and
(b) the closed-loop bandwidth.

5.35 An *RC* amplifier has a frequency response as shown in the graph of Fig. 5.18 overleaf.
(a) From this graph determine
    (i) the amplifier mid-frequency gain, and
    (ii) the amplifier bandwidth.
(b) Negative feedback is now applied to this amplifier such that 2% of the output is fed back in opposition to the input. Plot the closed-loop amplifier frequency response, and state
    (i) the new mid-frequency gain, and
    (ii) the new amplifier bandwidth.

5.36 An amplifier is to be designed to have a voltage gain factor of 30 ± 0.2%. The gain for the basic amplifier circuit is $A_v \pm 10\%$. Determine
(a) the open-loop gain A_v (hint: use simultaneous equations)
(b) the negative feedback fraction required to achieve this.

5.37 An amplifier has an open-loop gain of 50 dB and a closed-loop gain of 25 dB when negative feedback is applied. If the open-loop gain falls to 45 dB due to a faulty component, calculate the resulting closed-loop gain.

5.38 A voltage amplifier has 0.4% of its output fed back in antiphase to its input. Under this condition the voltage gain is 43.5 dB. Determine the open-loop gain.

5.39 An amplifier having an open-loop voltage gain factor of 250 is required to have a closed-loop gain of 63 by utilising negative feedback. This feedback is to be derived by a potential divider circuit connected across the output as shown in Fig. 5.19. Calculate
(a) the percentage of output voltage that must be fed back,
(b) the required value for R_2, and
(c) the modified value of output resistance if the open-loop value is 25 kΩ.

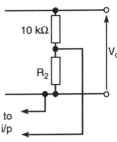

Fig. 5.19.

5.40 A 555 timer astable multivibrator circuit as shown on page 51 has $C_1 = 10$ nF, and $V_{cc} = +9$ V. It is required to obtain an output waveform having a frequency of approximately 70 Hz, with a mark to space ratio of 2:1. Determine suitable values for R_1 and R_2, and hence select the nearest preferred values.

5.41 An astable multivibrator circuit as in Fig. 5.20 is required to provide an output at a frequency of

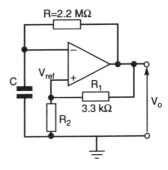

Fig. 5.20.

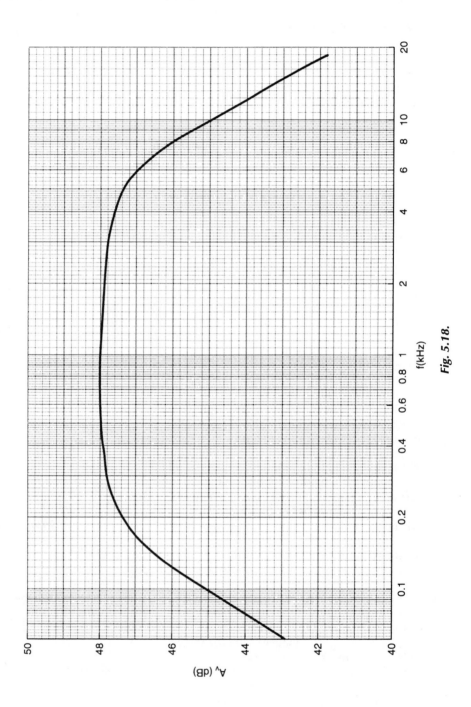

Fig. 5.18.

approximately 1 kHz. The supply for the circuit (not shown) is ± 15 V, and the amplifier saturation voltage is ± 13 V. Calculate

(a) the value of R_1 such that $V_{ref} = \pm 3.023$ V

(b) the value required for C, and hence select the nearest preferred value from those listed below

(c) the actual frequency output thus achieved, and

(d) the approximate output frequency obtained if R_2 is changed to 2.7 kΩ

390 pF, 470 pF, 560 pF, 680 pF.

5.42 A 555 timer monostable multivibrator circuit is illustrated on page 52. It is required to obtain an output pulse of duration as close as possible to 0.5 s. Resistors of 15 kΩ, 100 kΩ and 220 kΩ are available, as are capacitors of value 2.2 μF, 4.7 μF and 33 μF.

(a) Determine the most suitable combination of resistor and capacitor to be used as C_2 and R_2 in the circuit

(b) calculate the resulting percentage error in the output pulse duration

(c) it is not always necessary to include resistor R_1 and capacitor C_1. Explain the conditions under which they are required.

5.43 An a.c. voltage of $v = 5 \sin 200\pi t$ volt is to be converted into a squarewave of the same frequency by using a Schmitt trigger circuit as shown in Fig. 5.21. The power supply to the operational amplifier is ± 9 V, and its output saturation voltage is 2 V less than the power supply voltage. It is required that the switching action occurs 1.085 ms after each instant that the input waveform passes through zero.

(a) Determine the value to which R_2 must be set

(b) sketch, to scale, the input and output waveforms indicating all principal values on both the voltage and time axes

(c) if R_2 is adjusted to 6 kΩ, determine the instant of time for the input waveform at which the switching action will occur.

5.44 An operational amplifier is connected to passive components so as to perform as a monostable multivibrator, as shown on page 52. If R_2 is set to 1.8 kΩ, $R_3 = 10$ kΩ and $R = 680$ kΩ, determine the value of capacitor C that will result in an output pulse of duration 0.25 s.

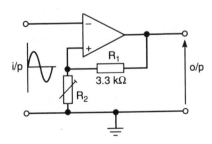

Fig. 5.21.

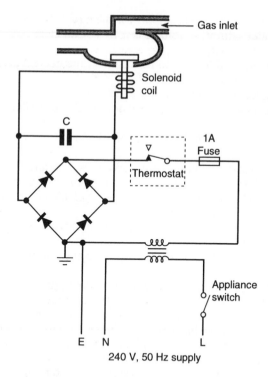

Fig. 5.22.

5.45 The flow of gas to a boiler is supplied via a fail-safe system consisting of a solenoid operated valve, which in turn is controlled by a thermostat switch, as shown in Fig. 5.22 above.

    When the current through the solenoid coil falls to 300 mA, the valve is returned to the closed position by a return spring (not shown). The following data applies to the system.

Solenoid valve coil:	operating voltage = 24 V
	resistance = 52 Ω
	inductance = 260 mH
	number of turns = 800
Solenoid magnet circuit:	effective cross-section = 0.75 cm^2
	total reluctance = 11.06 × 10^6 At/Wb
Diodes:	forward voltage drop = 0.6 V per diode
Transformer:	may be considered ideal (no losses)

Using the above data calculate
(a) the time taken for the solenoid to release the valve stem after the thermostat contact opens due to overheating in the boiler
(b) the transformer turns ratio required, assuming that the reservoir capacitor is sufficiently large to produce a ripple-free voltage to the solenoid, and
(c) the force exerted by the solenoid return spring in order to close the valve.

6 Electrical Machines

Equations

Transformers

Voltage ratio

$$\frac{V_p}{V_s} = \frac{N_p}{N_s}$$

Assuming an ideal transformer of efficiency 100%

Current ratio

$$\frac{I_p}{I_s} = \frac{N_s}{N_p}$$

Power ratio

$$P_o = P_i \text{ watt}$$

E.m.f. equation

$$E = 4.44\,\Phi_m Nf \text{ volt}$$

Iron loss (constant)

$$P_{Fe} = V_p I_o \cos\phi_o \text{ watt}$$

where I_o and ϕ_o represent the no-load primary current and phase angle

Copper loss (variable)

$$P_{Cu} = I_p^2 R_p + I_s^2 R_s \text{ watt}$$

Efficiency

$$\eta = \frac{V_s I_s \cos\phi}{V_s I_s \cos\phi + P_{Fe} + P_{Cu}} \text{ watt}$$

Maximum efficiency

This occurs when

$$P_{Cu} = P_{Fe}$$

$$\text{or, } \eta = \frac{\text{output}}{\text{output} + P_{Fe}}$$

$$x^2 P_{Cu} = P_{Fe}$$

where x is the fraction of rated kVA at which maximum efficiency occurs.

Equivalent resistance referred to the primary

$$R_e = R_p + R_s(N_p/N_s)^2 \text{ ohm}$$

Equivalent reactance referred to the primary

$$X_e = X_p + X_s(N_p/N_s)^2 \text{ ohm}$$

Equivalent impedance referred to the primary

$$Z_e = \sqrt{R_e^2 + X_e^2} \text{ ohm}$$

Alternators

Frequency generated $\qquad f = np$ hertz

where p is the number of pole **pairs** and n is speed in rev/s

E.m.f. generated per phase $\qquad E_{ph} = 2.22\ \Phi Zf$ volt

where $\qquad Z$ = number of useful conductors per phase

$\qquad \Phi$ = useful flux per pole

and assumes that both the distribution factor and pitch factor are equal to 1.0.

Power flow diagram

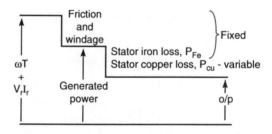

Three-phase induction motor

Stator 'speed' (speed of rotating field)

$$n_s = f/p \text{ rev/s}$$

Percentage slip

$$s = \frac{n_s - n_r}{n_s} \times 100\%$$

Power flow diagram

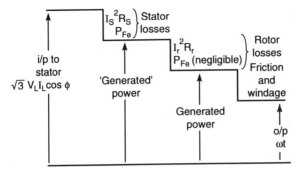

Three-phase synchronous motor

$$n_r = n_s = f/p \text{ rev/s (speed of rotating field)}$$

Power flow diagram

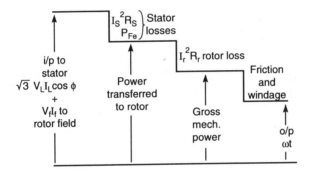

Direct current generators

E.m.f. equation $\qquad\qquad E = \dfrac{2p\Phi Zn}{a} \text{ volt}$

where

p = the number of pole **pairs**

Φ = useful flux per pole

Z = number of armature conductors

n = speed in rev/s

a = number of parallel paths through armature

Note: For a wave winding, $a = 2$; and for a lap winding $a = 2p$

Separately excited generator

Terminal voltage $V = E - I_a R_a$ volt

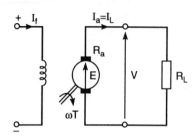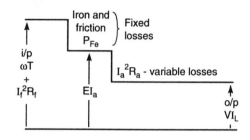

Shunt generator

Terminal voltage $V = E - I_a R_a$ volt

and $I_a = I_L + I_f$ amp

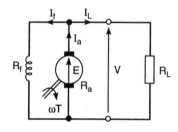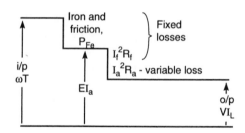

Series generator

Terminal voltage $V = E - I_a(R_a + R_f)$ volt

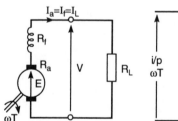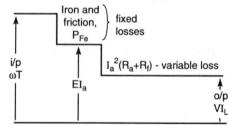

Compound generator (short shunt)

Terminal voltage $V = E - (I_a R_a + I_L R_{Se})$ volt

and $I_a = I_L + I_f$ amp

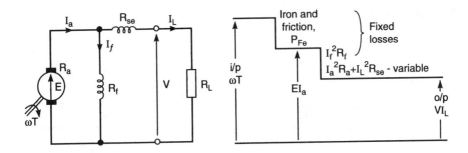

❑ **Note:** For all generators, η_{max} occurs when variable losses = fixed losses

Direct current motors

Speed 'equation'
$$\eta \propto \frac{E_b}{\phi} \text{ or } \omega \propto \frac{E_b}{\phi}$$

Torque 'equation'
$$T \propto \Phi I_a$$

Shunt motor

$I_L = I_a + I_f$ amp ('constant' flux machine); and $E_b = V - I_a R_a$ volt

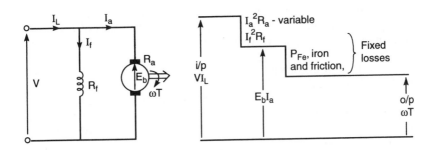

Series motor

$I_L = I_a = I_f$ amp (variable flux machine); and $E_b = V - I_a(R_a + R_f)$ volt

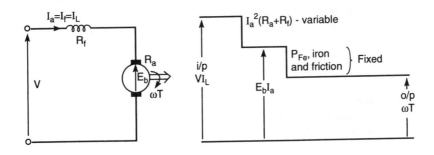

Compound motor (long shunt)

$$I_L = I_a + I_f \text{ amp}, \quad E_b = V - I_a(R_a + R_{Se}) \text{ volt}$$

When running light or on no-load, the input power to a motor equals the fixed losses, i.e. on no-load;

$$VI_L = P_{Fe} + I^2R_f \text{ watt for both shunt and compound motors}$$

$$VI_L = P_{Fe} \text{ watt for a series motor}$$

Maximum efficiency occurs when the variable losses = the fixed losses.

Assignment Questions _____

6.1 A single-phase transformer used with a 50 Hz supply has a laminated core of cross-section 0.05 m^2 which carries a maximum flux density of 1.2 T. Given that the number of primary and secondary turns are 240 and 80, respectively, determine
(a) the voltage transformation ratio,
(b) the primary and secondary voltages, and
(c) the power input when supplying a 55 A load at a power factor of 0.65.

6.2 A single-phase transformer has a turns ratio of 1:5. The primary is connected to a 230 V supply and a 500 Ω resistive load is connected across the secondary winding. Calculate
(a) the secondary voltage and current,
(b) the primary current, and
(c) the power delivered to the load.

6.3 A single-phase, 50 Hz transformer is required to provide a secondary voltage of 240 V when connected to a 6.6 kV supply. The transformer core has a cross-section measuring 20 cm × 20 cm. Determine a suitable number of primary and secondary turns if the core flux density is not to exceed 1.4 T.

6.4 A 250 kVA, 3300 V/250 V, 50 Hz transformer has 50 turns on its secondary winding. Calculate
(a) the primary and secondary currents on full-load,
(b) the maximum value of the core flux, and
(c) the number of primary turns.

6.5 A single-phase transformer rated at 23 kVA is connected to a 230 V, 50 Hz supply. There are 450 primary turns and 30 secondary turns. Calculate
(a) the secondary voltage on no-load

(b) the values for the primary and secondary current when supplying its rated full-load current to a resistive load, and

(c) the maximum value of the core flux.

6.6 The design data for a 50 Hz, single-phase isolating transformer is as follows:

$$\text{core cross-section} = 0.0095 \text{ m}^2$$
$$\text{peak core flux density} = 1.35 \text{ T}$$
$$\text{conductor diameter} = 1.5 \text{ mm}$$
$$\text{maximum current density in conductors} = 2.4 \text{ MA/m}^2$$
$$\text{primary supply voltage} = 230 \text{ V}$$

Using this data calculate

(a) the transformer kVA rating, and

(b) the approximate number of turns required on the primary winding.

6.7 (a) A 240 V, 50 Hz, single-phase, 2:1 step-down transformer draws a primary current of 1.2 A at a power factor of 0.42 lagging when the secondary winding is on open-circuit. Sketch the no-load phasor diagram and calculate the value for the transformer iron loss.

(b) If this transformer is now used to supply a 65 A load at a power factor of 0.85 lagging, sketch the on-load phasor diagram and determine the value and power factor of the resulting primary current. Any voltage drops due to the windings may be considered negligible.

6.8 A 4:1 step-down transformer draws a no-load primary current of 0.5 A at a lagging power factor of 0.2. Determine the primary current that would flow when the secondary is supplying a 30 A load at a power factor of 0.8 lagging.

6.9 The no-load primary current of a transformer is 3.5 A at a power factor of 0.22, when the primary is connected to a 230 V, 50 Hz supply. If the primary has 250 turns, calculate

(a) the peak value of the core flux,

(b) the magnetising component of the primary current, and

(c) the value for the transformer fixed losses.

6.10 The open and short-circuit tests on a 415 V/230 V single-phase transformer yielded the following results:

Open-circuit test: $\quad V_p = 415 \text{ V}$
$\qquad\qquad\qquad\quad I_o = 0.361 \text{ A}$
$\qquad\qquad\qquad\quad P_i = 30 \text{ W}$

Short-circuit test: $\quad V_p = 12 \text{ V}$
$\qquad\qquad\qquad\quad I_s = 20 \text{ A (full-load current value)}$
$\qquad\qquad\qquad\quad P_i = 65 \text{ W}$

Using these results determine

(a) the no-load primary phase angle

(b) the transformer efficiency when supplying full-load at power factor 0.8, and

(c) the transformer efficiency when supplying half full-load at power factor 0.8.

6.11 A three-phase, 50 Hz transformer has its primary winding delta connected and its secondary winding star connected. The number of turns per phase on the primary is five times that on the secondary, and the secondary line voltage is 415 V. A balanced load of 25 kW and power factor 0.75 is connected across the secondary terminals. Assuming no losses, calculate

(a) the primary voltage
(b) the phase and line currents in the secondary circuit, and
(c) the phase and line currents in the primary circuit.

6.12 The windings of a single-phase, 6:1 step-down transformer have the following values of resistance and leakage reactance:

$$R_p = 0.8\ \Omega; \qquad X_p = 5\ \Omega$$

$$R_s = 0.015\ \Omega; \quad X_s = 0.09\ \Omega$$

Calculate the primary voltage that needs to be applied in order to circulate a current of 120 A through the secondary circuit when the secondary terminals are short-circuited.

6.13 A single-phase, 1:4 step-up transformer has an equivalent resistance of 0.2 Ω and equivalent leakage reactance of 0.45 Ω, both being referred to the primary. Determine the secondary voltage developed when the primary is connected to a 230 V, 50 Hz supply, and a load of resistance 180 Ω and reactance 90 Ω is connected across the secondary terminals. The primary magnetising current may be considered negligible.

6.14 A 50 kVA transformer has an iron loss of 480 W and a full-load copper loss of 900 W. If the load power factor is 0.8, calculate
(a) the full-load efficiency
(b) the load at which maximum efficiency occurs, and
(c) the value of the maximum efficiency.

6.15 A three-phase, delta-connected motor has a power factor of 0.8 and an efficiency of 80% when providing an output of 90 kW. This motor is supplied from a delta/star connected transformer, the delta-connected primary of which is connected to a 600 V, 50 Hz, three-phase supply. The transformer has a turns ratio of 1.5:1, and may be considered loss-free. Sketch the circuit diagram and calculate
(a) the line voltage applied to the motor stator
(b) the line current supplied to the motor stator
(c) the current flowing in the motor stator winding, and
(d) the current drawn from the 600 V supply.

6.16 An eight-pole, single-phase alternator is required to generate 300 V r.m.s. at a frequency of 50 Hz. Given that the useful flux/pole is 25 mWb, calculate
(a) the number of stator conductors required, and
(b) the driving speed, measured in radian/second.

6.17 A three-phase, 50 Hz alternator has a star-connected stator winding. There are 50 stator slots, each containing 3 conductors. The machine has four rotor poles, and the useful flux/pole is 0.25 Wb. Calculate
(a) the output line voltage, and
(b) the speed of rotation, in rev/min.

6.18 A 10 MVA, 11 kV, 50 Hz, three-phase, star-connected alternator is driven at 300 rev/min. The stator winding is contained in 360 slots with six conductors per slot. Calculate
(a) the number of rotor poles, and
(b) the flux/pole required to provide the 11 kV line voltage on open circuit.

6.19 A 45 kVA, three-phase, 50 Hz alternator having a star-connected stator winding supplies a balanced three-phase load with a full-load current of 80 A at a power factor of 0.75 lagging. Under this condition the fixed losses are 3.5 kW and the efficiency is 83%. The d.c. power supplied to the rotor field winding is 150 W. Determine

(a) the mechanical power input, and
(b) the stator copper losses.

6.20 A 25 kVA, single-phase alternator supplies a full-load current of 125 A at a power factor of 0.8. Under this condition the iron, friction and windage loss is 0.9 kW, and the stator copper loss is 2.5 kW. The rotor field winding is connected to a 100 V d.c. supply and draws a current of 4 A. Determine

(a) the full-load efficiency
(b) the load at which maximum efficiency occurs
(c) the value for the maximum efficiency, and
(d) the input driving torque under this condition if the speed of rotation is 50 rev/s.

6.21 A workshop requires three 115 V, 400 Hz, single-phase supplies, each capable of providing a maximum current of 15 A at unity power factor. These three supplies may be arranged in a three-phase configuration, but must be independent of the normal mains supply. Equipment available are an alternator driven by a diesel engine, and a transformer, the details of which are given below

Alternator:	number of phases	3
	stator connection	star
	number of rotor poles	32
	active conductors/phase	260
	maximum allowable phase current	9 A
	efficiency	85%

Transformer:	number of phases	3
	rating	7.5 kVA
	primary winding	522 turns/phase, delta
	secondary winding	150 turns/phase, star
	efficiency	98%

Sketch a circuit diagram showing how this equipment should be arranged to provide the required workshop supplies, and hence calculate

(a) the alternator phase current
(b) the alternator phase voltage
(c) the speed required of the diesel engine
(d) the alternator flux/pole
(e) the actual kVA loading on the transformer when supplying the maximum load demanded, and
(f) the power output from the diesel engine.

6.22 A three-phase synchronous motor has 12 poles and operates from a 400 V, 50 Hz, three-phase supply. When on load the motor draws a line current of 110 A at a leading power factor of 0.8, and has an efficiency of 90%. Determine

(a) the speed of rotation
(b) the input power, and
(c) the output torque developed.

6.23 A two-pole, three-phase, 50 Hz induction motor runs with a slip of 3.5% when under load. Calculate its speed of rotation.

6.24 A three-phase induction motor is operated from a 415 V, 50 Hz three-phase supply. With no load connected to its shaft it runs at a speed of 1497 rev/min, with a slip of 0.2%. When on full-load it runs with a slip of 4%. Calculate
(a) the number of poles, and
(b) the full-load speed.

6.25 A four-pole, three-phase induction motor has a delta-connected stator winding and a cage rotor. When brake tested by the manufacturer the following results were obtained under full-load conditions

speed	1440 rev/min
line current	50 A
line voltage	400 V
supply frequency	50 Hz
input power, using the two-wattmeter method	$P_1 = 10.139$ kW $P_2 = 20$ kW
net brake force	680 N
effective brake diameter	500 mm

Using the above data, calculate
(a) the full-load power factor
(b) the full-load efficiency, and
(c) the full-load slip.

6.26 A 415 V, three-phase, four pole induction motor has a cage rotor and a delta-connected stator winding. It is supplied from a three-phase transformer, the primary of which is connected to a 6.6 kV, 50 Hz supply. When the motor is operating at its maximum possible efficiency its variable losses amount to 3.5 kW. Under this condition the output torque was measured as 746 N m, at a shaft speed of 1480 rev/min, the input power factor being 0.82 lagging. The circuit arrangement is shown in Fig. 6.1. Assuming a transformer efficiency of 100%, calculate
(a) the motor input power and efficiency
(b) the percentage slip
(c) the motor line current
(d) the transformer turns ratio, and
(e) the current drawn from the 6.6 kV supply.

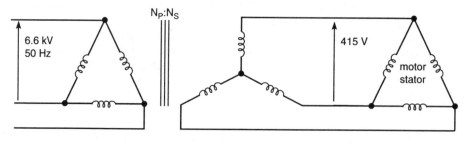

Fig. 6.1.

6.27 A four-pole d.c. generator has its armature wound with 500 conductors. When driven at a certain speed the e.m.f. generated in each conductor is 1.2 V, and the current flowing through each conductor is 50 A. Determine the total machine e.m.f. and generated power if the winding is (a) lap wound, and (b) wave wound.

6.28 An eight pole d.c. machine has 480 lap-connected armature conductors. Calculate the useful flux per pole required to generate an e.m.f. of 250 V, when driven at 1000 rev/min.

6.29 A six-pole d.c. machine has a wave-connected armature of 420 conductors, and the useful flux/pole is 25 mWb. Determine the speed at which it must be driven in order to generate an e.m.f. of 450 V.

6.30 The armature resistance of a six-pole d.c. shunt generator is 0.02 Ω. The armature is lap-wound with 560 conductors. When driven at 800 rev/min the machine produces an armature current of 500 A at a terminal voltage of 240 V. Calculate the useful flux per pole.

6.31 A d.c. shunt generator has a field winding resistance of 185 Ω. When the machine is supplying an output of 90 kW the terminal voltage is 450 V, and the generated e.m.f. is 460 V. Calculate
(a) the armature resistance, and
(b) the value of generated e.m.f. when supplying 50 kW at a terminal voltage of 480 V.

6.32 The short-shunt compound generator shown in Fig. 6.2 has armature, shunt field and series field resistances of 0.6 Ω, 100 Ω and 0.4 Ω respectively. When supplying a load of 7.5 kW at a terminal voltage of 250 V, the power supplied by the driving motor is 10 kW. Determine
(a) the generated e.m.f.
(b) the efficiency
(c) the iron and friction loss, and
(d) the total fixed losses.

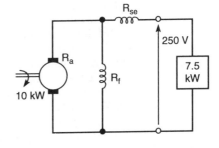

Fig. 6.2.

6.33 A d.c. motor draws an armature current of 100 A from a 500 V supply. The armature resistance is 0.22 Ω. The machine has four poles and the armature is lap wound with 800 conductors. The flux per pole is 0.045 Wb. Calculate
(a) the speed of rotation
(b) the gross torque developed by the armature, and
(c) the shaft output torque if the friction and windage loss is 2.5 kW.

6.34 The series motor shown in Fig. 6.3 runs at 550 rev/min when drawing a current of 100 A from a 220 V supply. The resistances of the armature and field windings are 0.14 Ω and 0.02 Ω, respectively. Determine the motor speed when the current drawn from the 220 V supply is 45 A, given that the useful flux per pole at 100 A is 0.02 Wb and that at 45 A is 0.012 Wb.

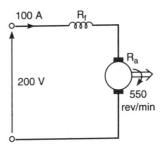

Fig. 6.3.

6.35 The short-shunt compound generator shown in Fig. 6.4 supplies a 12.5 kW load at a terminal voltage of 250 V. The resistance of the armature, shunt field and series field windings are 0.1 Ω, 200 Ω and 0.15 Ω, respectively. The iron, friction and windage losses total 500 W. Determine

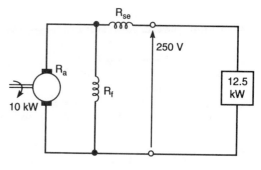

Fig. 6.4.

(a) the input power and the efficiency

(b) the shaft driving torque if the speed of rotation is 1100 rev/min.

6.36 The armature of a 240 V shunt motor rotates at 12.5 rev/s and carries a current of 6 A when on no-load. The armature resistance is 0.5 Ω. Determine

(a) the speed when loaded and carrying an armature current of 80 A, and

(b) the speed under the conditions of (a) above if, due to armature reaction, the flux is weakened to 98.5% of its unloaded value.

6.37 A shunt generator has a full-load output of 15 kW at a terminal voltage of 250 V. The armature and shunt field resistances are 0.6 Ω and 125 Ω, respectively, the friction and windage loss being 0.6 kW. Calculate

(a) the input driving power required on full-load, and the resulting efficiency

(b) the output current at which maximum efficiency occurs, and

(c) the value of the maximum efficiency.

6.38 A 200 V shunt motor draws a line current of 2.2 A from a 200 V supply when running light. The resistances of the armature and shunt field windings are 0.25 Ω and 180 Ω, respectively.

(a) Calculate the motor output and efficiency when the current drawn from the supply is 30 A.

(b) If this motor is now driven as a generator, calculate the input power and efficiency when it is supplying a 30 A load at a terminal voltage of 200 V.

6.39 A 150 kW, 240 V shunt generator was tested, yielding the following results

(i) When running light as a motor, at full speed, the line current drawn from a 240 V supply was 32 A, and the field current was 10 A.

(ii) With the machine stationary, an applied voltage of 5 V across the armature caused a current of 350 A to flow.

Using these results determine the machine efficiency at (a) full-load, and (b) half full-load. You may assume the flux remains constant.

6.40 A three-phase induction motor is used to drive a d.c. shunt generator as shown in Fig. 6.5. System details are as follows:

Induction motor:		
	number of poles	2
	supply voltage	415 V, 50 Hz
	input line current	22 A
	power factor	0.85
	percentage slip	4%
	efficiency	86%

Shunt generator:		
	armature resistance	0.5 Ω
	field resistance	300 Ω
	load supplied	9 kW at 250 V

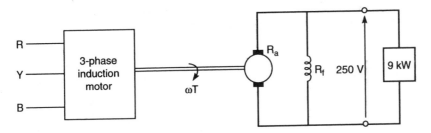

Fig. 6.5.

For the conditions given determine
(a) the input torque to the d.c. generator
(b) the generator armature copper loss and efficiency
(c) the power supplied to the d.c. load when the generator is operating at its
    maximum possible efficiency. You may assume that the flux remains constant.

6.41 A small-scale generating plant is arranged as in Fig. 6.6. The d.c. exciter is used
to supply the rotor field winding of the three-phase alternator.

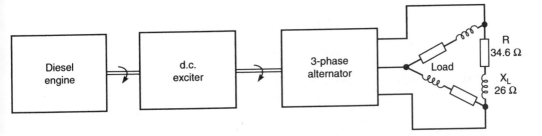

Fig. 6.6.

Details of the electrical machines are as follows:

Alternator:	stator winding	star-connected
	line output voltage	600 V
	frequency	50 Hz
	efficiency	90%
	stator conductors/phase	30
	reluctance of magnetic circuit	1.442×10^4 At/Wb
	number of rotor poles	4
	number of turns on rotor winding	150
Exciter:	output voltage	110 V
	efficiency	80%

Using the above data, determine
(a) the power output from the alternator
(b) the alternator flux required
(c) the output power from the exciter
(d) the speed of the diesel engine, and
(e) the shaft torque developed by the diesel engine

6.42 A three-phase induction motor is supplied from the secondary winding of a three-phase step-down transformer, the primary of which is connected to an 11 kV, 50 Hz supply. The circuit arrangement is shown in Fig. 6.7.

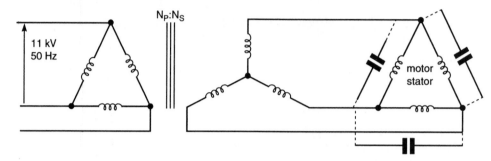

Fig. 6.7.

Details of the two machines are as follows:

Transformer:	number of primary turns	6000 (delta-connected)
	number of secondary turns	131 (star-connected)
	supply voltage	11 kV
	frequency	50 Hz
	efficiencymay be assumed	100%

Induction motor:	number of poles	8
	full-load rating	100 kVA
	power factor	0.85 lagging
	full-load efficiency	85%
	full-load slip	3%

Using this data, determine
(a) the phase voltage applied to the stator of the induction motor
(b) the full-load speed
(c) the motor output torque
(d) the motor stator line and phase currents
(e) the line current drawn from the 11 kV supply, and
(f) the kVAr rating of each phase of a delta-connected capacitor bank connected in parallel with the motor stator in order to improve the power factor to unity.

6.43 A motor-generator set consisting of a d.c. shunt motor and a single-phase alternator is illustrated in Fig. 6.8 opposite. Details of the two machines are as follows.

D.C. motor:	supply voltage	110 V
	field resistance	80 Ω
	armature resistance	0.5 Ω
	iron and friction loss	425 W

| **Alternator:** | full-load rating | 2.4 kVA |
| | full-load output voltage | 240 V |

frequency	50 Hz
number of poles	2
number of stator conductors	1500
number of rotor conductors	250
reluctance of magnetic circuit	1.39×10^5 At/Wb

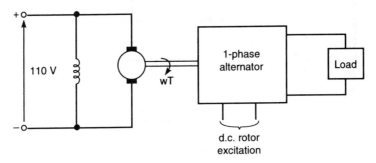

Fig. 6.8.

When the alternator is supplying a purely resistive load at its full rated output, the d.c. motor draws a line current of 34 A from the 110 V supply. For this condition, and using the data provided, calculate
(a) the motor efficiency
(b) the alternator efficiency
(c) the d.c. excitation current required by the alternator field winding, and
(d) the motor output torque.

6.44 A separately excited d.c. generator is used to charge a bank of 50 cells that are connected in series with each other. The circuit is illustrated in Fig. 6.9, and incorporates an automatic cut-out arrangement which ensures that the generator is not connected to the battery bank until the generator terminal voltage exceeds the total e.m.f. of the cells by a certain amount. The generator then remains connected all the time it is charging the cells, but is automatically disconnected should the output current be reversed. Data for the circuit is as follows:

Generator:	armature resistance	0.5 Ω
	field power input	176 W
	iron and friction loss	55 W
Cells:	each of e.m.f.	2.0 V
	each of internal resistance	0.02 Ω
Cut-out circuit:	resistor R_1	200 Ω
	resistor R_2	2 kΩ
	comparator	µA 741

The circuit operation is as follows. The contactor closes when the p.d. across its coil exceeds 8 V. The electronic comparator (C) switches to an output of +10 V when the potential at input 'x' rise above that at input 'y', and remains in that state until the potential at 'x' falls to a value of 10.5 V less than that at 'y'. The comparator then switches to an output of −10 V, and remains in this state until the potential at 'x' once more rises to equal that at 'y'. The characteristics for the comparator are shown in Fig. 6.10.

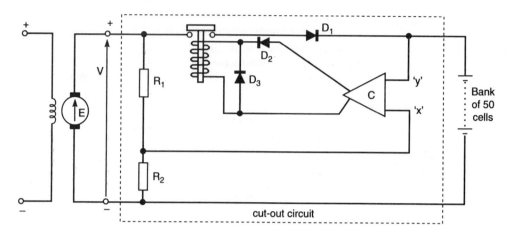

Fig. 6.9.

Assuming that the forward voltage drop of all diodes is negligible, determine
- (a) the generator input driving power and efficiency when the contactor contacts are closed and the generator is supplying a charging current of 15 A
- (b) the value of generator terminal voltage required to *just* cause the contactor contacts to close
- (c) the potential of 'x' relative to 'y' (i.e. the p.d. between 'x' and 'y') after the contacts close, assuming no change in generated e.m.f. from the value required in (b) above, and that the current has reached a steady value, and
- (d) Explain (i) why the comparator 'switch-on' and 'switch-off' voltage input conditions need to differ, and (ii) the functions of the diodes D_1, D_2 and D_3.

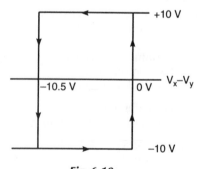

Fig. 6.10.

7 Control Systems and Logic

System Elements and Analogies

Energy storage and dissipative elements

System	Storage elements		Dissipative elements
	'K.E.'	'P.E.'	
Electrical	inductor, L; $W = \frac{1}{2}Li^2$	capacitor, C; $W = \frac{1}{2}CV^2$	resistor, R; $R = V/i$
Linear mechanical	mass, m; $W = \frac{1}{2}mv^2$	spring, k; $W = \frac{1}{2}ks^2$	damper, R; $R = F/v$
Rotary mechanical	*polar mass moment of J; $W = \frac{1}{2}J\omega^2$ inertia,	torsion bar, or k; $W = \frac{1}{2}k\theta^2$ spring,	damper, R; $R = T/\omega$

$*J = mr^2$ kg m^2, where r = radius of gyration.

System variables

System	Analogous variables			
Electrical	voltage, V	current, i	charge, q	rate of change of current, $\dfrac{di}{dt} = \dfrac{d^2q}{dt^2}$
Linear mechanical	force, F	velocity, v	displacement, s	acceleration, $a = \dfrac{d^2s}{dt^2}$
Rotary mechanical	torque, T	angular velocity, ω	angular displacement, θ	angular acceleration, $\alpha = \dfrac{d^2\theta}{dt^2}$

System element equations

System	'Inertial' term	'Stiffness' term	'Damping' term
Electrical	inductor voltage $V = L\dfrac{d^2q}{dt^2}$	capacitor voltage $V = \dfrac{q}{C}$	resistor voltage $V = R\dfrac{dq}{dt}$
Linear mechanical	inertia force $F = m\dfrac{d^2s}{dt^2}$	stiffness force $F = ks$	damping force $F = R\dfrac{ds}{dt}$
Rotary mechanical	inertia torque $F = J\dfrac{d^2\theta}{dt^2}$	stiffness torque $T = k\theta$	damping torque $F = R\dfrac{d\theta}{dt}$

❑ **Note:** Capacitance, $C \equiv 1/k$ (reciprocal of spring stiffness coefficient) In general, a series electrical system is analogous to a parallel mechanical system.

A gearbox behaves in a **similar** manner to an electrical transformer, **but** note the following:

$\omega_i T_i = \omega_o T_o$ (assuming no losses)

$$\frac{\omega_o}{\omega_i} = \frac{1}{n}$$

$$\frac{T_o}{T_i} = \frac{n}{1}$$

First-order systems (response to step-input)

General equation

$$Y = A\frac{dx}{dt} + Bx$$

where

Y = step-input disturbance

x = system variable

A = coefficient of first derivative term

B = coefficient of 'constant' term

System response to the step-input disturbance is a transient exponential change, with a time constant, $\tau = A/B$ seconds.

At any time t seconds after the disturbance, the variable will have a value given by

$$x = \frac{Y}{B}(1 - e^{-t/\tau}) \qquad \text{for exponential growth}$$

$$\text{or, } x = \frac{Y}{B}e^{-t/\tau} \qquad \text{for exponential decay}$$

Second-order systems (response to step-input)

General equation

$$Y = a\ddot{x} + b\dot{x} + cx$$

$$\text{or, } \frac{Y}{a} = \ddot{x} + \frac{b\dot{x}}{a} + \frac{cx}{a}$$

$$\text{or, } \frac{Y}{a} = \ddot{x} + 2\zeta\omega_n\dot{x} + \omega_n^2 x$$

This last equation is most frequently used in analysing control systems, where

$$Y = \text{step-input disturbance}$$

$$x = \text{system variable}$$

$$\ddot{x} = \frac{d^2x}{dt^2}$$

$$\dot{x} = \frac{dx}{dt}$$

$$a, b \text{ and } c = \text{the system element coefficients} \\ \text{(e.g. } L, m, k, R, \text{ etc.)}$$

$$\zeta = \text{system damping ratio} = \frac{b}{2\sqrt{ac}}$$

$$\omega_n = \text{system natural frequency, in rad/s } = \frac{c}{\sqrt{a}}$$

if $b^2 < 4ac$ then $\zeta < 1$; system is underdamped

$b^2 = 4ac$ then $\zeta = 1$; system is critically damped

$b^2 > 4ac$ then $\zeta > 1$; system is overdamped

System response to a step-input disturbance is an exponentially decaying sinusoidal oscillation, given by the expression:

$$x = Ae^{-\alpha t} \sin \omega_d t$$

where $Ae^{-\alpha t}$ = exponentially decaying amplitude

and ω_d = system damped frequency of oscillation.

Damping ratio

$$\zeta = \frac{R}{R_{crit}}$$

where R_{crit} is the damping coefficient which *just* prevents any oscillation.

Damped frequency

$$\omega_d = \omega_n \sqrt{1 - \zeta^2} \text{ rad/s}$$

$$f_n = f_n \sqrt{1 - \zeta^2} \text{ hertz}$$

First-order systems (response to a forcing function)

When the input disturbance to a first-order system is alternating (and of any waveshape), the system will respond as either a differentiator or an integrator, depending upon the actual storage element contained. See Chapter 5 for problems involving *C-R* and *L-R* circuits.

Second-order systems (response to a forcing function)

The following applies when the forcing function is *sinusoidal*:

1. If the input frequency is very low, then the system 'displacement' will closely follow that of the input, in both amplitude and phase.
2. As the frequency is increased, so the amplitude and phase lag of the system displacement will increase. The system amplitude will reach its maximum when the input frequency equals the natural frequency of the system. At this particular frequency the system displacement will lag the input by 90°. Also, if the system damping is low, then the resonance effect will result in the system amplitude exceeding that of the input. In mechanical systems this condition is normally avoided, since severe mechanical damage can result.
3. At input frequencies greater than the system natural frequency, the system displacement amplitude will rapidly decrease, whilst the phase lag will continue to increase, tending towards 180°.

The above effects are illustrated in the two figures opposite.

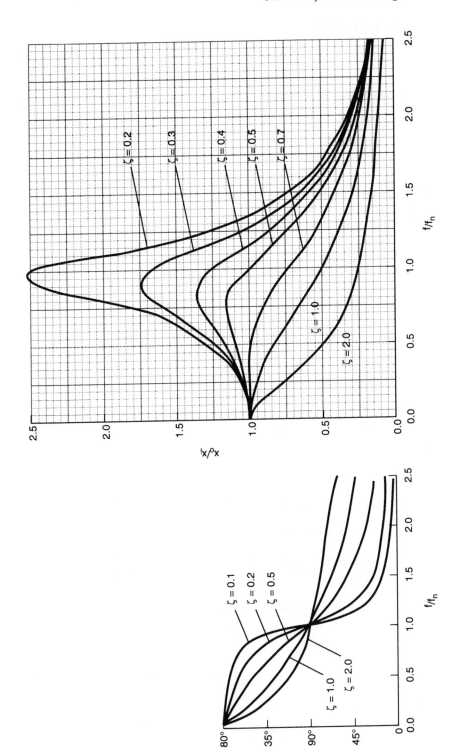

Feedback strategies

Proportional (positional) feedback: provides the basic closed-loop path for the feedback of output data to the system comparator.

Derivative (velocity) feedback: enables simulated damping in order to improve system stability.

Integral feedback: provided to reduce steady-state error, but can also promote instability.

Operational amplifiers used for system simulation

Inverter

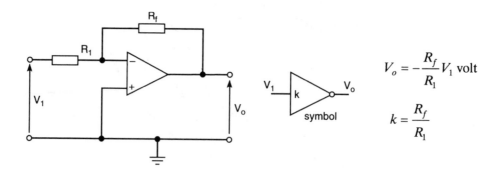

$$V_o = -\frac{R_f}{R_1}V_1 \text{ volt}$$

$$k = \frac{R_f}{R_1}$$

Summer

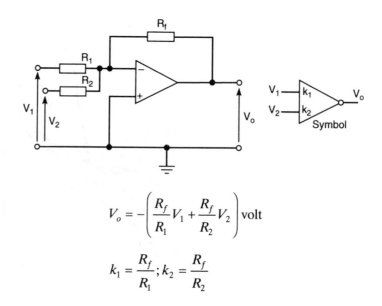

$$V_o = -\left(\frac{R_f}{R_1}V_1 + \frac{R_f}{R_2}V_2\right)\text{ volt}$$

$$k_1 = \frac{R_f}{R_1}; k_2 = \frac{R_f}{R_2}$$

Integrator

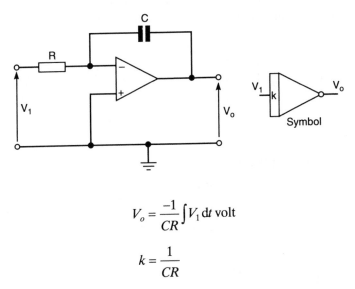

$$V_o = \frac{-1}{CR} \int V_1 \, dt \text{ volt}$$

$$k = \frac{1}{CR}$$

Differentiator

This amplifier configuration is normally avoided in analogue computer (simulator) circuits because it greatly magnifies any noise present.

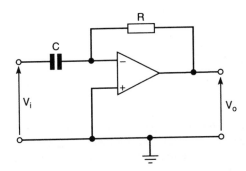

$$V_o = -CR \frac{dV_1}{dt} \text{ volt}$$

Differential amplifier

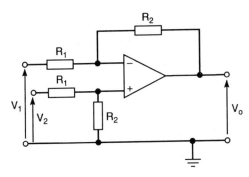

$$V_o = \frac{R_2}{R_1}(V_2 - V_1) \text{ volt}$$

Coefficient potentiometer

This is a precision potentiometer, and must be adjusted to provide the desired coefficient only when connected to its load in order to eliminate loading errors

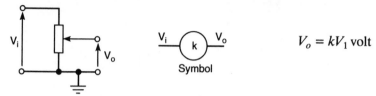

Symbol

$$V_o = kV_1 \text{ volt}$$

Combinational logic

Basic rules of Boolean algebra

$$A + A = A \qquad A.A = A$$
$$A + 1 = 1 \qquad A.1 = A$$
$$A + 0 = A \qquad A.0 = 0$$
$$A + \bar{A} = 1 \qquad A.\bar{A} = 0$$
$$A(B + C) = A.B + A.C$$
$$A.B + B.C = B(A + C)$$

DeMorgan's Law

$$\overline{A.B} = \bar{A} + \bar{B}$$
$$\overline{A.\bar{B}} = \bar{A} + B$$
$$\overline{\bar{A}.\bar{B}} = A + B$$
$$A.B = \overline{\bar{A} + \bar{B}}$$

Logic functions and gates

❏ **Note:** The British Standard (B.S.) symbols for the logic gates will be used in this text, with the exception of the inverter, for which the alternative symbol will be used throughout. However, it will be found that industry tends to use only the alternative symbols.

NOT function or inverter

A — [1] — $F = \bar{A}$ A — ▷ — $F = \bar{A}$

B.S.   Alternative

OR function

A — [≥1] — $F = A + B$ A — ⟩ — $F = A + B$
B —   B —

B.S.   Alternative

AND function

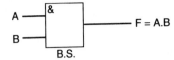

A ─┤& ├─ F = A.B
B ─┤ ├
   B.S.

A ─┤ ⟩─ F = A.B
B ─┤
Alternative

NOR function

A ─┤≥1 ├o─ F = $\overline{A+B}$
B ─┤   ├
   B.S.

A ─┤ ⟩o─ F = $\overline{A+B}$
B ─┤
Alternative

NAND function

A ─┤& ├o─ F = $\overline{A.B}$
B ─┤ ├
   B.S.

A ─┤ ⟩o─ F = $\overline{A.B}$
B ─┤
Alternative

Exclusive OR function

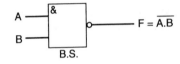 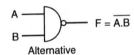

A ─┤=1 ├o─ $F = A.\overline{B} + \overline{A}.B$
B ─┤   ├ $= (A+B).\overline{A.B}$
   B.S.

A ─┤ ⟩)o─ $F = A.\overline{B} + \overline{A}.B$
B ─┤ $= (A+B).\overline{A.B}$
Alternative

Simple latch (memory) circuits using universal gates

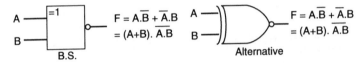

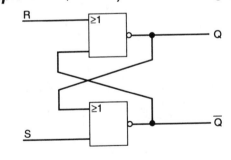

Truth Table

S	R	Q	$\overline{Q}$
0	0	no change	
1	0	1	0
0	1	0	1
1	1	disallowed	

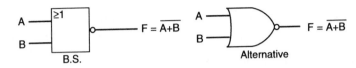

Assignment Questions _____

7.1   A 5 kg mass moves with a velocity of 10 m/s. If all of the kinetic energy stored by this mass is to be transferred to a spring by causing it to compress by 10 cm, determine the spring stiffness coefficient.

7.2   A torsion bar is subjected to an overall twist of 0.1 radian. Determine the torsional stiffness if under this condition the energy stored is 2.5 kJ.

7.3   A mass of 3 kg is placed on the horizontal platform of a weighing machine, which has a spring of stiffness 250 N/m and a damper of damping coefficient 620 Ns/m. Assuming that the mass of the platform is negligible and that the acceleration due to gravity is 9.81 m/s^2, calculate
  (a) the total displacement of the platform,
  (b) the displacement 5 s after the mass is placed on the platform, and
  (c) the initial velocity.

7.4   A first-order system possesses a time constant of 10.5 s. Calculate the time taken for it to reach a state corresponding to 50% of its final steady state following a step-input disturbance.

7.5   A first-order mechanical system is illustrated in Fig. 7.1. This system may be considered to have negligible mass. Determine the approximate time taken for the system to reach its new steady state after a step input force (F) of 49.05 N is applied to it.

7.6   A logic gate circuit is shown in Fig. 7.2. For this circuit write down the Boolean expression for the function F, and implement this function using NOR gates only.
  You may assume complements are available.

7.7   The block diagram for a control system is given in Fig. 7.3. Determine the overall transfer function for this system.

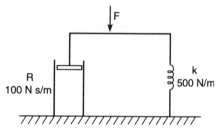

Fig. 7.1.

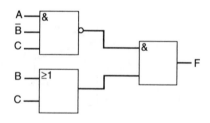

Fig. 7.2.

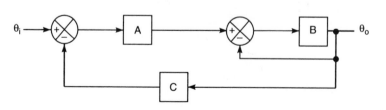

Fig. 7.3.

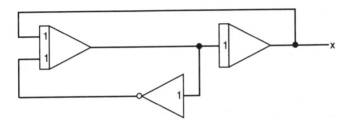

Fig. 7.4.

7.8 The block diagram of an analogue simulator circuit is shown in Fig. 7.4. Determine the system differential equation being simulated.

7.9 The circuit of Fig. 7.5 is subjected to a sudden step-input voltage of E volts when the switch is closed. Determine
   (a) the system differential equation, and
   (b) the value of resistance that will just prevent any oscillations (the critical damping).

7.10 The second-order mechanical system illustrated in Fig. 7.6 has a mass m of 24 kg, a damper coefficient, R of 24 Ns/m and a spring stiffness coefficient, k of 10 N/m. Calculate the increase of damping coefficient required to provide critical damping.

7.11 Calculate the values of resistance and inductance of the circuit shown in Fig. 7.7 such that the system has an undamped natural frequency of 10 Hz and a damping ratio of 0.5.

7.12 A simple first-order mechanical system consists of a spring of stiffness 10 N/m which is mounted in parallel with a damper of damping coefficient 20 Ns/m. This system is subjected to a step-input force of 5 N which extends the spring. Determine for this system
   (a) the time constant
   (b) the final extension of the spring, and
   (c) the time taken for the spring to extend by 0.1 m.

7.13 For the electrical system shown in Fig. 7.8 calculate
   (a) the undamped natural frequency
   (b) the damping ratio, and
   (c) the damped frequency of oscillation when subjected to a step-input disturbance.

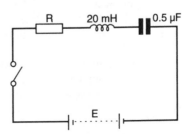

Fig. 7.5.

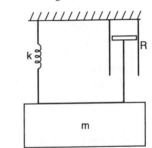

Fig. 7.6.

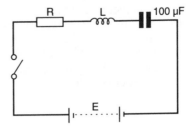

Fig. 7.7.

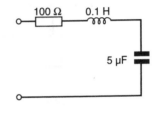

Fig. 7.8.

7.14 A 10 mV step-input is applied to the input of the operational amplifier circuit shown in Fig. 7.9. Calculate the value of the output voltage, V_o, 0.5 s after the input is applied.

7.15 An 8 kg mass is suspended from a helical spring of stiffness 840 N/m, and its movement is damped by a dashpot of damping coefficient 130 Ns/m. Determine
  (a) the system damping ratio, and
  (b) the frequency of oscillations when subjected to a sudden step-input displacement.

7.16 The operational amplifier circuit of Fig. 7.10 has voltages of +5 V, −6 V and +1.5 V applied to input terminals A, B and C, respectively. Calculate the resulting output voltage obtained.

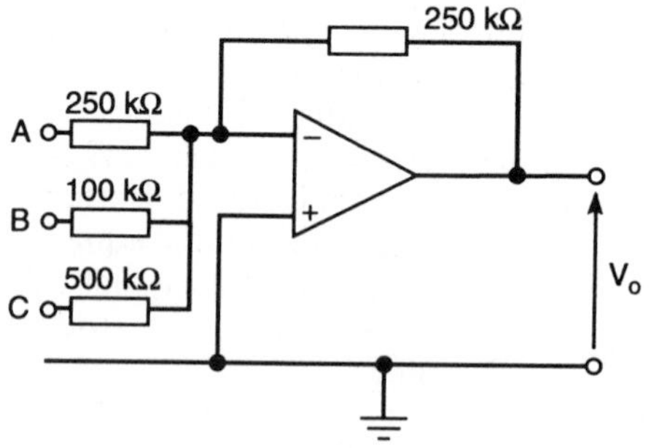

7.17 The truth table shown opposite gives the only states of four switches (ON ≡ 1; OFF ≡ 0) which provide an overall ON output for the function F. Using this table
  (a) form the Boolean expression for function F
  (b) reduce this expression to its simplest
  (c) sketch the simplest logic gate circuit that will implement the function using only NOR gates.

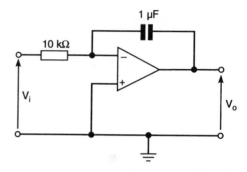

Fig. 7.9.

Fig. 7.10.

A	B	C	D	F
0	1	0	1	1
0	1	1	1	1
1	1	0	1	1
1	1	1	1	1

7.18 Calculate the frequency of oscillation of an undamped mechanical system consisting of a 3 kg mass suspended from a spring of stiffness 120 N/m.

7.19 The block diagram of an analogue computer (simulator)circuit is illustrated in Fig. 7.11. Derive the system differential equation and hence determine whether the system would be over, under or critically damped.

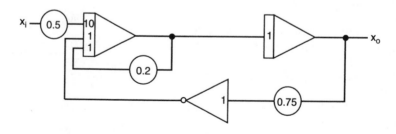

Fig. 7.11.

7.20 Sketch a logic gate circuit that will implement the logic function

$$F = A.B + \bar{A}.\bar{B} + A.C$$

using (a) AND, OR and NOT gates, and (b) NAND gates only.

7.21 The pan and spring assembly of a household weighing machine has a total mass of 0.1 kg. When a mass of 4 kg is placed on the pan the spring is compressed 25 mm. Determine the spring stiffness coefficient and the undamped natural frequency of oscillation for the system. Take the acceleration due to gravity as 9.81 m/s.

7.22 The response of a second-order system to a step-input disturbance is described by the following differential equation.

$$8 = 4\ddot{\theta} + 6\dot{\theta} + 4\theta$$

Using this equation, determine
(a) the natural undamped frequency
(b) the damping ratio, and
(c) the damped frequency of oscillation.

7.23 Sketch the logic gate circuit, using no more than three 2-input NAND gates, that will implement the logic function represented by the truth table shown opposite.

A	B	C	F
0	0	0	0
1	0	0	0
0	1	0	0
1	1	0	1
0	0	1	0
1	0	1	0
0	1	1	1
1	1	1	1

7.24 A spring of stiffness 50 kN/m is fixed at one end and the movement of the free end is restricted by a dashpot providing viscous friction of 25 kNs/m. Compared with the other coefficients, the mass of the system may be considered negligible. If this system is subjected to a step input force of 4 kN,
(a) write down the system differential equation, and
(b) calculate the displacement of the spring 0.07 s after the input force is applied.

7.25 For the operational amplifier circuit shown in Fig. 7.12, determine its output voltage V_o 0.5 s after the input, V_i is set to a constant voltage of 3 V.

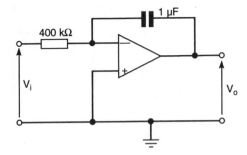

Fig. 7.12.

7.26 A damper of coefficient 10 Ns/m is connected between the two ends of a spring of stiffness 20 N/m. Write down the system differential equation and determine the time taken for the spring to displace by 0.15 m when subjected to a step-input force of 3.5 N.

7.27 A constant force of 150 N is applied to a mass of 20 kg which is at rest at time $t = 0$. The motion of the mass is resisted by viscous friction of 10 Ns/m. For this system
(a) deduce the differential equation
(b) calculate the steady-state velocity achieved, and
(c) sketch the equivalent electrical circuit, quoting the values of components used.

7.28 Using (a) the rules of Boolean algebra, and (b) Karnaugh map technique, simplify the following logic expressions

(i) $F = \overline{A}.\overline{B}.\overline{C} + \overline{A}.B.C + A.\overline{B}.\overline{C}$

(ii) $F = P.\overline{R} + \overline{P}.(Q + R) + P.Q.(R + \overline{Q})$

(iii) $F = A.\overline{B}.C + A.B.(\overline{A} + C) + B.C.(\overline{B} + A)$

7.29 Implement the following functions using the minimal number of
(a) NAND gates only, and (b) NOR gates only. Assume that complements are available.

(i) $F = A.B.\overline{C} + \overline{A}.B.C + A.\overline{B}.C$

(ii) $F = A.\overline{B} + \overline{A}.B$

(iii) $F = A.B.(A + \overline{B}.\overline{C}) + \overline{A}.\overline{B}.C$

(iv) $F = (P.\overline{Q} + \overline{P}.R).(\overline{P}.\overline{Q}.R)$

7.30 (a) Write down the Boolean expression for the logic function performed by the circuit shown in Fig. 7.13
(b) simplify the above expression and sketch the logic circuit that will implement the minimised expression.

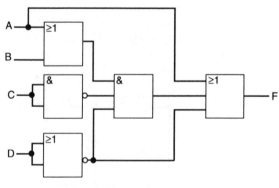

Fig. 7.13.

7.31 The logic circuit, using 7400 series i.c.s, for a sequencer is shown in Fig. 7.14 opposite. For this circuit
(a) Write down the logic expressions for the three outputs X, Y, and Z
(b) A fault develops such that outputs X and Y are normal, but output Z has some missing '1' states. Suggest which gate may be faulty, and the nature of this fault.

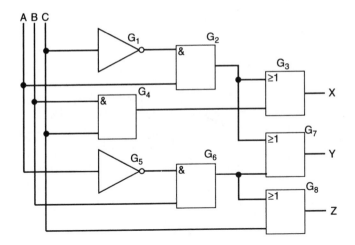

Fig. 7.14.

7.32 In an industrial process, a controlled volume of a liquid is to be heated. The general arrangement is illustrated in Fig. 7.15 below. The heating element (E) and the inlet and outlet valves (V_1 and V_2) are controlled by a logic circuit, using 7400 series i.c.s, as shown in Fig. 7.16 on the next page.

(a) Draw up a truth table showing the output functions V_1, V_2, and E for all possible combinations of the sensors H, T, and L, assuming that H, L, and T are active when in the high (logical '1') state

(b) state the likely effects when (i) G_3 is short-circuited between input and output, and (ii) G_1 input is open circuited.

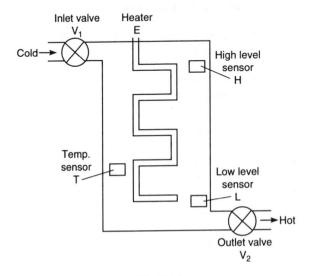

Fig. 7.15.

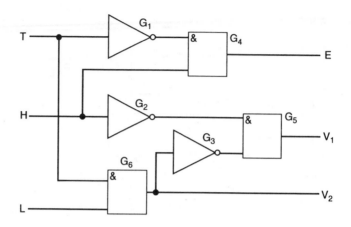

Fig. 7.16.

7.33 The d.c. supply for the operational amplifier shown in Fig. 7.17 is ± 10 V. Determine the output voltage when the input is (a) $+1.5$ mV, (b) -30 mV, and (c) $+200$ mV.

7.34 The operational amplifier circuit shown in Fig. 7.18 has an output voltage of $+5.25$ V when V_i is 6.8 V. Determine the corresponding value for input V_2.

7.35 Considering the electrical system shown in Fig. 7.19, and with the switch in position 'A'

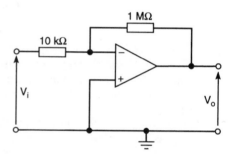

Fig. 7.17.

(a) Calculate the capacitor terminal voltage 4 ms after connection of the supply. You may assume that the capacitor is initially uncharged

(b) after the capacitor becomes fully charged the switch is moved to position 'B'. Under this condition, calculate

(i) the circuit damping ratio

(ii) the frequency of the resulting oscillations, and

(iii) the value to which R must be adjusted in order to achieve critical damping.

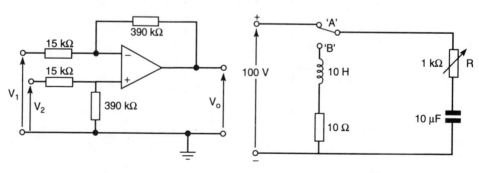

Fig. 7.18.   **Fig. 7.19.**

7.36 The response of a second-order rotational system, when subjected to a step-input disturbance, is described by the equation

$$\ddot{\theta}_o + 4\dot{\theta}_o + 8\theta_o = 0.8$$

Given that the polar mass moment of inertia (J) of the system is 0.5 kgm^2, determine
(a) the torsional stiffness and damping coefficients
(b) the system natural frequency of torsional oscillation, and
(c) the damping ratio and damped frequency of oscillation.

7.37 The response of a first-order system, when subjected to a step-input disturbance, is described by the equation

$$\dot{x} + 3x = 2$$

(a) Assuming a computer reference supply of ± 10 V, sketch the analogue computer circuit that will simulate this system response
(b) calculate the steady state value of x.

7.38 The design for a small-scale suspension system is to be tested by simulating its response to a step-input displacement on an analogue computer. The parameters for the system are as follows:

$$\text{mass}, m = 2.5 \text{ kg}$$

$$\text{damping coefficient}, R = 6.5 \text{ Ns/m}$$

$$\text{spring stiffness}, k = 7.5 \text{ N/m}$$

(a) Write down the system differential equation
(b) sketch the analogue computer circuit
(c) determine
  (i) the system natural frequency of oscillation
  (ii) the damping ratio
  (iii) the damped frequency of oscillation, and
(d) sketch a graph of the system response to the step-input.

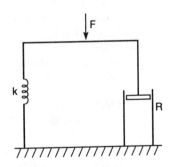

Fig. 7.20.

7.39 A spring–dashpot system is shown in Fig. 7.20, where the spring stiffness is 2 kN/m and the damping coefficient is 500 Ns/m.
(a) Sketch the equivalent electrical circuit, quoting the values of the components, and
(b) sketch the response of the system when subjected to a square wave input of frequency 3 Hz.

7.40 Figure 7.21 represents a spring balance in which a platform of negligible mass is supported by a spring of stiffness 1 kN/m and the vertical movement is controlled by a damper of damping coefficient 500 Ns/m.

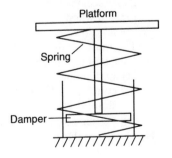

Fig. 7.21.

(a) Obtain an expression for the displacement of the platform if a downwards force F is suddenly applied to it at time $t = 0$

(b) if a mass of 5 kg is slid horizontally onto the platform and then released, determine the initial velocity and the final displacement

(c) obtain the system differential equation for the complete system including the 5 kg mass.

7.41 An analytical instrument of mass m is shown in idealised form in Fig. 7.22. It is known that when in use the structure will be subjected to sinusoidal vibrations. It is desired to minimise the effects of these vibrations on the instrument by introducing some form of antivibration mounting.

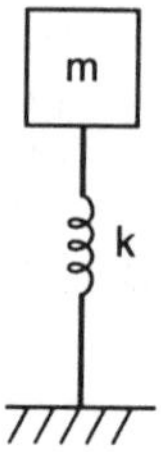

(a) Calculate the undamped natural frequency of the system and the damping coefficient required to ensure critical damping, given that $m = 5$ kg and $k = 20$ N/m

(b) explain why it is important to specify the lowest acceptable frequency of the vibrations, and

(c) sketch the approximate response of the undamped system if the vibrations are in the form of a square wave having a period of 2 s.

Fig. 7.22.

7.42 The angular position of a small radar antenna is controlled by means of a system which includes both positional and velocity feedback. The system utilises a d.c. drive motor which is coupled to the antenna via a speed reduction gearbox.

(a) Draw a block diagram of a suitable control system indicating the positional and velocity feedback loops, the types of transducers employed and their position within the system. Briefly explain the purpose and method of operation of each of the feedback loops

(b) fitted at various points of the above system are switches and cut-outs to prevent overload and allow safe maintenance. The logical requirements for these are specified in the table opposite. Derive a minimised logic circuit, using NAND gates only, to provide the required output control conditions.

Input conditions				Output
A	B	C	D	
0	0	0	0	1
1	0	0	0	1
0	1	0	1	1
1	1	0	1	1
0	1	1	1	1
1	1	1	1	1
0	0	1	0	1
1	0	1	0	1
all other conditions				0

(c) Explain what is meant by 'velocity lag' and its cause for a system such as that in part (a) above. How may the effects of velocity lag be reduced?

7.43 In an industrial process three gases, A, B, and C are to pass over a workpiece in a furnace, via a master valve. A safety requirement is that gas C must be present when gas B flows and also during standby periods when gases A and B are not flowing. Gas A may flow either with or without the presence of gas C. The master valve is to be controlled by a logic circuit employing NAND gates only, and it may be assumed that complements are available.

(a) Define the logic system variables

(b) construct a truth table to illustrate the solution to the control problem and simplify the resulting Boolean expression

(c) sketch a logic circuit that will implement the simplified expression.

7.44 The generalised block diagram of an electro-hydraulic control system is shown in Fig. 7.23 at the top of the next page.

(a) Derive the system closed-loop transfer function

(b) given that the element transfer functions are A = 10^5 m/V and B = 20 mV/m, calculate the output displacement when an input voltage of 1 mV, representing input demand s_i, is applied to the comparator, and

(c) calculate the change of output displacement if, due to a fault, the transfer function A is reduced to 10^3 m/V.

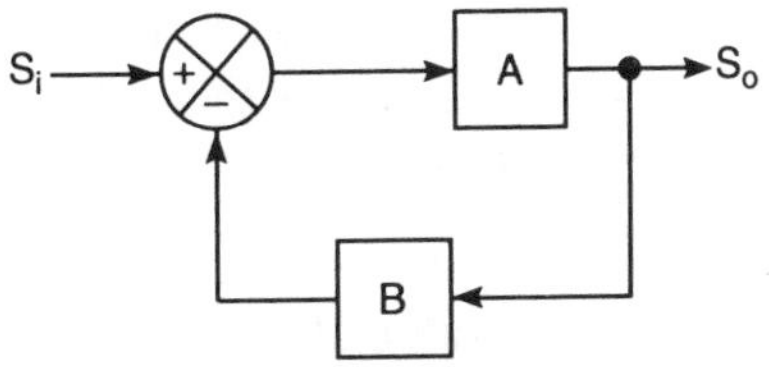

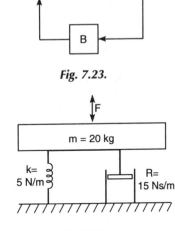

Fig. 7.23.

7.45 The mechanical system in Fig. 7.24 is subjected to a sinusoidal input force of F = 5 sin 2*t* newton. Determine

(a) the mechanical 'impedance'
(b) the maximum velocity
(c) the maximum amplitude of displacement
(d) the undamped natural frequency, and
(e) the damping ratio.

Fig. 7.24.

7.46 The logic conditions for the operation of an electrically driven lift are as follows. The motor will start only if:

(i) the lift doors are fully closed
(ii) the maximum load has not been exceeded, and
(iii) a floor select button has been pressed (momentary signal only).
The motor will stop when:
(iv) the emergency stop button is pressed (momentary signal only)
(v) the lift arrives at the selected floor.

(a) Define the variables for the logic control system and write down the corresponding Boolean expression which defines the 'motor run' conditions,

(b) draw a logic circuit diagram to implement the Boolean expression, utilising only 3-input NAND gates.

7.47 A solid cylindrical mass is attached to a shaft of torsional stiffness 7.5 MNm/rad. The mass has a density of 7000 kg/m³ and dimensions as shown in Fig. 7.25.

Write down the system differential equation which describes the response to a step-input displacement applied to the mass, and hence determine the frequency of oscillations so caused.

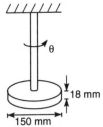

7.48 A flywheel consists of a horizontal mild steel disc of diameter 750 mm and thickness 22 mm. The density of the steel is 7900 kg/m³. This flywheel is attached to the end of a shaft of torsional stiffness 12 Nm/rad.

Fig. 7.25.

Determine

(a) the undamped natural frequency of oscillation
(b) the angular damping required to provide critical damping, and
(c) the damped frequency of oscillation if a damping ratio of 0.5 is applied.

7.49 A second-order rotary mechanical system is subjected to a step-input torque and its response is described by the differential equation

$$\ddot{\theta} + 4\dot{\theta} + 8\theta = 0.8$$

Given that the system inertia is 0.5 kgm^2, determine
(a) the values for the system torsional stiffness and the damping coefficient
(b) the natural frequency of oscillation
(c) the system damping ratio and the frequency of the damped oscillations, and
(d) the value of the applied input torque.

7.50 A positional control system is illustrated in Fig. 7.26. The angular position of the load is controlled by altering the setting on the input potentiometer. A tachogenerator is available to provide velocity feedback, the degree of this feedback being adjustable by means of another potentiometer.

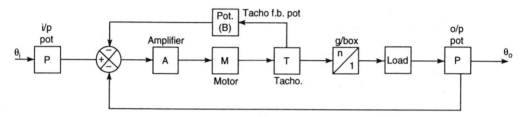

Fig. 7.26.

The parameters for the various system elements are as follows:

Amplifier gain, A	120 mA/V
Motor transfer function, M	0.02 Nm/mA
Load inertia, J_o	8 kgm^2
Motor inertia, J_m	0.01 kgm^2
Gearbox ratio, n	30:1
Motor viscous friction, R	0.055 Nms/rad
Tachogenerator transfer function, T	0.16 Vs/rad
Input and output potentiometers transfer function, P	1.2 V/rad
Load and gearbox viscous friction	negligible

(a) Derive the system differential equation that describes the response to a step-input demand of 0.8 radian, and hence calculate
(b) the system natural frequency of oscillation
(c) the damping ratio
(d) the steady state error for an input velocity demand of 3 rev/min
(e) the setting required (B) of the tachogenerator potentiometer in order to provide critical damping, and the resulting steady state error for the same ramp displacement input demand specified in (d) above.

Answers to Assignment Questions

Chapter 1

[1.1]   0.819 m

[1.2]   2.681 Ω

[1.3]   (a) 210.2 Ω (b) 1.9 A (c) 186.9 Ω

[1.4]   (a) 4.53 V (b) 30.46 A (Cu); 19.54 A (Al) (c) 138 W (Cu); 88.5 W (Al)

[1.5]   12 Ω; 0.75 W

[1.6]   (a) 1.767×10^{-8} Ωm (b) 2.17 Ω; 2.42 Ω

[1.7]   1.4 μV

[1.8]   10

[1.9]   11.46 A; 0.252 kWh

[1.10] (a) (i) 560 Ω, 1/8 W (ii) 1/4 W (b) 6.187 V (O.K.)

[1.11] (a) aluminium cable (b) 46.37 A

[1.12] (a) 270 Ω in series with two 1.5 kΩ resistors in parallel (b) 58.4 mW (270 Ω); 81.1 mW (in each 1.5 kΩ resistor)

[1.13] –20°C

[1.14] 2 mΩ

[1.15] 1.102 kΩ

[1.16] (a) R_1 = 2.2 kΩ; R_2 = 330 Ω (b) 19.9:1 (c) 34.8 mW (R_1); 4.73 mW (R_2)

[1.17] (a) 12.5 A (10.5 V batt.); 7.5 A (10 V batt.) (b) 9.375 kJ; 6.75 kJ

[1.18] (a) 5 V (b) 2.5 kΩ (c) 10 mW (R_C); 2 mW (R_E)

[1.19] (a) 2 kΩ (b) $V_{BC} = V_{DC} = 1.667$ V (c) 15.76 μA from D to B

[1.20] 6 mV (B positive w.r.t. A)

[1.21] (a) 22 kΩ (b) (i)1.167 V (ii) 1.031 V

[1.22] (a)V_A = 393 V; V_B = 391 V; V_C = 389.5 V (b) 1.61 kW

[1.23] 105 A (A to B); 5 A (B to C); 70 A (A to C)   (b) V_B = 489.5 V;
V_C = 488.8 V
[1.24] 10.25 V; 52.5 mW

[1.25] (a) 2.528 V/rad (b) 6.319 V (c) 6.189 V (d) –2.06%

[1.26] (b) (i) Cct.1, 270 Ω; 0.3 W; 4.2 p: Cct.2, 120 Ω; 1 W; 25.5 p:
Cct.3, 27 Ω; 7 W; 95 p (ii) circuit 2

[1.27] 4.7 kΩ (to ensure saturation)

[1.28] (b) V_{GS} = –1.5 V; I_D = 4.58 mA; V_{DS} = 13 V (c) 66.7 kΩ; 3.3 mS (d) 7.5

[1.29] (a) 1.25 kΩ (b) 680 Ω (calculated value 693 Ω) (c) V_{CE} = 6.1 V;
I_C = 8.7 mA; I_B =120 μA (d) 100 kΩ (calc. value 94.2 kΩ) (e) 76.2
(f) 5.7 kΩ (g) A_i = 66.25; A_V = 36.5

[1.30] 0.533; 1.6

[1.31] (a) 20 mA (b) 95 μA (c) 0.722 V (d) 218 Ω

[1.32] (a) 2.75 kΩ (b) 32 lux (c) 3/4 f.s.d.

[1.33] (a) V_B = 234 V; V_C = 223 V; V_D = V_E = 219 V (b) V_B = 237.5 V;
V_C = 235.25 V; V_D = 238.25 V (c) ring distributor – higher consumer
voltages and smaller cable cross-section

[1.34] (a) 12 000 rev/min (b) four complete cycles in 5 ms (c) 3.963 V (d) 39 kΩ

Chapter 2 _____

[2.1] (a) 450 μJ (b) 6 s

[2.2] (a) 87.44 V (b) 4.1 N attraction (c) 109.3 kV/m

[2.3] (a) 815 pF; 81.5 pF (b) 7.34 nC (c) 33 nJ

[2.4] (a) 4.7 μF (b) 9.4 μF (c) 1.77 μF (d) 1.97 μF (e) 14.7 μF (f) 62 μJ

[2.5] (a) failure due to dielectric breakdown (b) borderline (c) satisfactory

[2.6] (a) 10.625 μC (b) 4.5 cm (c) 53.8 pF

[2.8] R_1 and R_2 – potential divider; R_E – bias stabilisation; R_C – develops o/p voltage; C_1 and C_2 – a.c. coupling; C_3 – a.c. bypass to preserve d.c. bias under a.c. signal conditions (b) C_3 o/c resulting in negative feedback

[2.9] (b) At low frequencies reactance of C_1 and C_2 each act as part of potential divider circuits, thus attenuating input to transistor and output to load. At high frequencies, stray circuit capacitance and internal capacitance of transistor have a shunting effect on the output

[2.10] 10.85 pF; 1.81 pF

Chapter 3

[3.1]   (a) 270 mm^2 (b) 597 × 10^3 At/m (c) 23.58 × 10^6 At/Wb

[3.2]   (a) 82.76 At/m; 104 μT (b) 1.082 N

[3.3] 291 mV

[3.4]   (a) 169.8 (b) 896 μWb

[3.5]   (c) (i) 102.5 mT (ii) 162.5 At/m (iii) 3.3 A

[3.6]   (a) 120 μWb (b) 14.4 mA (c) 2653 (d) 12.5 H (e) 45 V

[3.7]   (a) 184.2 mA (b) 2.3 × 10^6 At/Wb (c) 0.978 H

[3.8]   (a) 204 turns (b) 8.16 m (c) 328 Ω (d) 14.93 At (e) 24.45 nH
(f) 1.43 μH

[3.9]   4.8 A

[3.10] (a) 0.218 A

[3.11] (a) B_A = 0.816 T; B_B = 2.45 T; B_C = B_g = 0.49 T (b) 69.4 mH
(c) 112.3 mH

[3.12] (a) (i) 0.122 μN (ii) 3.653 Ω (iii) 66.67 kΩ

[3.13] (a) L_A = 176.2 mH; L_B = 44 mH (b) 303.85 mH (aiding);
136.55 mH (opposing) (c) series opposition

[3.14] (a) 0–5 ms: e_1 = –60 V, e_2 = –38 V; 5–11 ms: e_1 = e_2 = 0 V;
11–19 ms: e_1 = +75 V, e_2 = +47 V (b) N_1 = 2472; N_2 = 1888

[3.15] 69 turns

[3.16] 28 A

[3.17] (a) 1082.5 At (b) 6928 (c) 14.82 mm

Chapter 4

[4.1] 16.65 mA; 23.55 mA

[4.2] 1.91 A; 2.12 A

[4.3] (a) (i) 83.3 Hz; 11 V; 7.78 V (ii) 16.7 kHz; 1.1 V; 0.778 V (iii) 1.67 kHz; 0.22 V; 0.156 V (b) (i) 11 sin 166.7πt volt (ii) 1.1 sin 33.3 × 10^3πt volt (iii) 0.22 sin 3.33 × 10^3πt volt

[4.4] (a) V_1 = 8.84 V; V_2 = 5.3 V (b) 84° or 1.47 rad (c) 15.23 sin (6.66 × 10^3πt – 0.512) volt

[4.5] 87.57 sin (628t + 0.257) mA

[4.6] 0.1 H, 50 Ω; 470 nF

[4.7] Box 1: inductor 50 mH, 75 Ω; Box 2: resistor 50 Ω; Box 3: capacitor 0.22 μF

[4.8] 5 μF; 55 mH, 19.82 Ω

[4.9] (a) 50 Ω (b) 0.184 H, 81.64 Ω (c) 0.916; 337 W

[4.10] 34.4 μF

[4.11] (a) 7.44 A (b) 0.956 (c) 1.635 kW

[4.12] (a) I_1 = 6.976 A; I_2 = 6.118 A; I_3 = 2.857 A (b) I = 8.375 A (c) 1.896 kW

[4.13] 38.97 Ω; 23.4 μF

[4.14] (a) 50 Hz (b) 12.94 Ω; 0.154 H (c) 203 W

[4.15] (a) 423.85 Hz; 17.75; 23.85 Hz (b) V_1 = 10 V; V_2 = 532.6 V; V_3 = 533 V

[4.16] (a) 50 (b) 20 nF (c) 48 Ω

[4.17] (a) I_1 = 1.9 A; I_2 = 1.59 A (b) 0.512 A (c) 97.4 W (d) 26 μF

[4.18] (a) 50 (b) 8.5 nF (c) 50 Ω, 53.1 mH (d) 36 V

[4.19] (a) 58.57 Hz (b) 266 Ω (c) 37.6 mA

[4.20] (a) 34.8 pF (b) 22.86 kΩ

[4.21] (a) 10 kHz (b) 9.88 V (c) 5 V

[4.22] 19.8 A; 0.99 lagging; 4.51 kW; 650 VAr

[4.23] 39 μF

[4.24] (a) 37.3 μF (b) 16.8 μF

[4.25] (a) 68 kW; 87.46 kVA (b) 26 kVAr

[4.26] 17.46 kVA; 0.916 lagging

[4.27] 45 mV

[4.28] (a) 25 mV (b) 508 μW

[4.29] (a) 6 kΩ (calc), so ,choose 6.8 kΩ (b) 1.7 mW

[4.30] (a) $I_{ph} = I_L = 4.96$ A; 1.846 kW
(b) $I_{ph} = 8.59$ A; $I_L = 14.88$ A; 5.537 kW

[4.31] (a) $I_{ph} = I_L = 25$ A (b) $I_{ph} = 1.67$ A; $I_L = 2.89$ A (c) 19.5 kW

[4.32] (a) 123.2 Ω, 0.294 H (b) 1.5 A; 19.5 kW

[4.33] (a) $V_p = 1.905$ kV; $V_s = 500$ V
(b) primary: $I_{ph} = I_L = 87.43$ A; secondary: $I_{ph} = 333.3$ A; $I_L = 577.4$ A

[4.34] (a) 1.5 kW (b) 0.822 (c) $I_{ph} = 1.52$ A; $I_L = 2.63$ A

[4.35] (a) 31.1 A (b) 17.96 A (c) 12.35 kW; 4.89 kW

[3.36] (a) $I_R = 34.64$ A; $I_Y = 28.87$ A; $I_B = 20.21$ A (b) $I_N = 12.35$ A

[4.37] (a) 17.32 A (b) 33.66 kW

[4.38] (a) $I_{RY} = 3.2$ A; $I_{YB} = 8$ A; $I_{BR} = 10$ A
(b) $I_R = 12.87$ A; $I_Y = 10.89$ A; $I_B = 9.17$ A

Chapter 5

[5.1] $E_o = 44$ V; $R_o = 40$ kΩ

[5.2] $I_{sc} = 22.5$ μA; $R_o = 8$ kΩ

[5.3] 156.7 mV ($h_{ie} = 1.8$ kΩ); 165.4 mV ($h_{ie} = 2.4$ kΩ); 171 mV ($h_{ie} = 3$ kΩ)

[5.4] (a) (i) 67.6 mV; 12.3 mA (ii) 104.2 mV; 10.4 mA (iii) 133.6 mV; 8.9 mA
(iv) 155.65 mV;7.78 mA (b) 20 Ω; 1.21 mW (d) zero (balanced bridge)

[5.5] −12 dB

[5.6] 179 mV

[5.7] (a) +8 dB (b) 40 mV

[5.8] 2.7 μW

[5.9] 49.2 dB

[5.10] (a) 0.5 μW; 3.16 μW; 0.5 mW (b) −33 dBm; −25 dBm; −3 dBm

[5.11] (a) 253 mW (b) 24 dBm

[5.12] 14.2 dB

[5.13] 14.14 V

[5.14] 31.35 dB; 21.35 dB

[5.15] (a) 221 μs (cap); 106.4 μs (ind) (b) 0.836 mA (cap); 4.99 mA (ind)
(c) 9.5 μJ (cap); 6.2 μJ (ind)

[5.16] (a) 20.9 mA/s (b) 24.9 V (c) 96.7 ms (d) 363 ms

[5.17] (a) 4 ms (b) 0.253 mA (c) 6.4 mJ (d) 20 ms (e) 0.219 A

[5.18] (a) 0.48 A (b) 2.1 ms (c) 160 mA

[5.19] 1 μF

[5.21] (a) 1.064 V; 1.036 V; −2.6% (b) 2.128 V; 2.018 V; −5.2%
(c) 8.511 V; 6.932 V; −18.6%

[5.22] 7.5 s

[5.23] 0.6 sin (10t + 3π/4) volt

[5.24] (a) 0.12 V pk-pk (b) 0.05%

[5.25] (a) 14.56 V (b) −4.7 dB

[5.26] (a) 0.554 V (b) 12.57 V (c) −27 dB

[5.27] $R_1 = 1149\ \Omega$; $R_2 = 756\ \Omega$

[5.28] (a) 320 Ω (b) 20 V (c) −6 dB

[5.29] (a) 25 (b) 2.7%

[5.30] 1.25%

[5.31] 4.4 dB

[5.32] (a) 22.23 (b) 24.46 (c) 20.21

[5.33] (a) 0.002 (b) $R_{out} = 0.19\ \Omega$; $R_{in} = 80\ M\Omega$

[5.34] (a) 0.056 (b) 56.23 kHz

[5.35] (a) (i) 48 dB (ii) 9.9 kHz (b) (i) 32.4 dB (ii) 60.2 kHz

[5.36] (a) 151.5 (b) 0.027

[5.37] 24.64 dB

[5.38] 51.43 dB

[5.39] (a) 0.0119 (b) 120 Ω (c) 6.3 kΩ

[5.40] $R_1 = R_2 = 690\ k\Omega$ (choose 680 kΩ)

[5.41] (a) 1 kΩ (b) 480 pF (choose 470 pF) (c) 1.02 kHz (d) 500 Hz

[5.42] (a) 4.7 μF and 100 kΩ gives $T = 0.517$ s (b) + 3.4%

[5.43] (a) 2.7 kΩ (c) 1.79 ms after passing through zero

[5.44] 2.2 μF

[5.45] (a) 2.35 ms (b) 9.52:1 (c) 2.5 N

Chapter 6

[6.1] (a) 3:1 (b) E_p = 3.197 kV; E_s = 1.066 kV (c) 38.1 kW

[6.2] (a) 1150 V; 2.3 A (b) 11.5 A (c) 2.645 kW

[6.3] N_p = 550; N_s = 20

[6.4] (a) I_p = 75.76 A; I_s = 1000 A (b) 22.5 mWb (c) 660 turns

[6.5] (a) 15.33 V (b) I_p = 100 A; I_s = 1500 A (c) 2.3 mWb

[6.6] (a) 0.975 kVA (b) 81 turns

[6.7] (a) 121 W (b) 33.51 A; 0.84 lagging

[6.8] 7.88 A

[6.9] (a) 4.14 mWb (b) 3.414 A (c) 177.1 W

[6.10] (a) –78.45° (b) 95.1% (c) 97.55%

[6.11] (a) 1198 V (b) I_{ph} = I_L = 46.37 A (c) I_{ph} = 9.27 A; I_L = 16.06 A

[6.12] 164.8 V

[6.13] 892.75 V

[6.14] (a) 96.67% (b) 29.21 kW (c) 96.81%

[6.15] (a) 693 V (b) 117.2 A (c) 67.66 A (d) 135.32 A

[6.16] (a) 108 (b) 78.54 rad/s

[6.17] (a) 2.4 kV (b) 1500 rev/min

[6.18] (a) 20 poles (b) 79.5 mWb

[6.19] (a) 40.51 kW (b) 3.413 kW

[6.20] (a) 84% (b) 12 kW (c) 84.5% (d) 43.93 Nm

[6.21] (a) 7.47 A (b) 231 V (c) 1500 rev/min (d) 1 mWb (e) 5.175 kVA
(f) 6.212 kW

[6.22] (a) 500 rev/min (b) 60.97 kW (c) 1048 Nm

[6.23] 2895 rev/min

[6.24] (a) 1500 rev/min (b) 1440 rev/min

[6.25] (a) 0.87 (b) 85.1% (c) 4%

[6.26] (a) 122.62 kW; 94.3% (b) 1.33% (c) 208 A (d) 27.55:1 (e) 13.08 A

[6.27] (a) 150 V; 30 kW (b) 300 V; 30 kW

[6.28] 31.25 mWb

[6.29] 857 rev/min

[6.30] 33.48 mWb

[6.31] (a) 0.0494 Ω (b) 485.3 V

[6.32] (a) 281.6 V (b) 75% (c) 814 W (d) 1.5 kW

[6.33] (a) 797 rev/min (b) 573 Nm (c) 543 Nm

[6.34] 956 rev/min

[6.35] (a) 89.5% (b) 121.3 Nm

[6.36] (a) 633 rev/min (b) 643 rev/min

[6.37] (a) 18.41 kW; 81.5% (b) 42.82 A (c) 82.9%

[6.38] (a) 89.18% (b) 89.8%

[6.39] (a) 91.77% (b) 89.1%

[6.40] (a) 38.33 Nm (b) 77.86%; 678.4 W (c) 15.34 kW

[6.41] (a) 24.95 kW (b) 0.104 Wb (c) 1.1 kW (d) 1500 rev/min (e) 185.3 Nm

[6.42] (a) 415 V (b) 727.5 rev/min (c) 1116 Nm (d) $I_L = 163.7$ A; $I_{ph} = 94.5$ A (e) 6.2 A (f) 20.66 kVAr

[6.43] (a) 70.4% (b) 91.2% (c) 0.8 A (d) 8.38 Nm

[6.44] (a) 2005 W; 84.25% (b) 110 V (c) 10 V

Chapter 7

[7.1] 50 kN/m

[7.2] 0.5 MNm/rad

[7.3] (a) 117.7 mm (b) 46 mm (c) 47.5 mm/s

[7.4] 7.28 s

[7.5] 1 s

[7.6] 3 × NORs: G_1 i/ps $\overline{A}$, B, $\overline{C}$; G_2 i/ps B, C; o/ps from G_1 and G_2 input to G_3

[7.7] AB/(1 + B + ABC)

[7.8] $\ddot{x} - \dot{x} - x = 0$

[7.9] (a) $E = L\ddot{q} + R\dot{q} + q/C$ (b) 400 Ω

[7.10] 6.984 Ns/m

[7.11] 159.2 Ω; 2.53 H

[7.12] (a) 2 s (b) 0.5 m (c) 0.446 s

[7.13] (a) 225 Hz (b) 0.354 (c) 210.5 Hz

[7.14] −0.5 V

[7.15] (a) 0.793 (b) 0.994 Hz

[7.16] +9.25 V

[7.17] (a) $F = \overline{A}.B.\overline{C}.D + \overline{A}.B.C.D + A.B.\overline{C}.D + A.B.C.D$ (b) $F = B.D$
(c) Two NORs as inverters with i/ps B and D, and o/ps of these to final NOR gate

[7.18] 1 Hz

[7.19] $5x_i = \ddot{x}_o + 0.2\dot{x}_o + 0.75x_o$; underdamped ($\zeta = 0.116$)

[7.21] 1.57 kN/m; 3.11 Hz

[7.22] (a) 0.159 Hz (b) 0.75 (c) 0.105 Hz

[7.23] 3 × NANDs: G_1 i/ps A,B; G_2 i/ps B,C; G_3 i/ps being o/ps from G_1 and G_2

[7.24] (a) $4 = 25\dot{s} + 50s$ (b) 10.45 mm

[7.25] −3.75 V

[7.26] $3.5 = 10\dot{s} + 20s$; 0.28 s

[7.27] (a) $150 = 20\dot{s} + 10\dot{s}$ (b) 15 m/s (c) 10 Ω resistor in series with 20 H inductor

[7.28] (a) $F = \overline{B}.\overline{C} + \overline{A}.B.C$ (b) $F = Q + \overline{P}.R + P.\overline{R}$ (c) $F = A.C$

[7.29] (a) (i) $4 \times$ NAND (ii) $3 \times$ NAND (iii) $3 \times$ NAND (iv) $3 \times$ NAND
(b) (i) $5 \times$ NOR (ii) $4 \times$ NOR (iii) $4 \times$ NOR (iv) $2 \times$ NOR

[7.30] (a) $F = A + \overline{D} + \overline{C}.\overline{D}.(A + B)$ (b) $F = A + \overline{D}$

[7.31] (a) $X = A.\overline{C} + B.C$; $Y = A.\overline{C} + \overline{A}.B$; $Z = \overline{A}.B + C$

[7.33] (a) −150 mV (b) +3 V (c) approx. −10 V (saturation)

[7.34] +7 V

[7.35] (a) 32.97 V (b) (i) 0.505 (ii) 15.92 Hz (iii) 1.99 kΩ

[7.36] (a) $R = 2$ Nms/rad; $k = 4$ Nm/rad (b) 0.45 Hz (c) 0.707; 0.318 Hz

[7.37] $x = 0.667$

[7.38] (a) $F = 2.5\ddot{s} + 6.5\dot{s} + 7.5s$ (c) (i) 0.276 Hz (ii) 0.751 (iii) 0.182 Hz

[7.39] (a) R of 500 Ω in series with C of 500 μF

[7.40] (a) $s = F(1 - e^{-t/0.5})/1000$ m (b) 49.05 mm; 0.0981 m/s
(c) $49.05 = 5\ddot{s} + 500\dot{s} + 10^3 s$

[7.41] (a) 0.318 Hz; 20 N/m

[7.42] (b) $F = A.D + \overline{B}.\overline{D}$; yields $5 \times$ NANDs (two as inverters)

[7.43] (b) $F = A.\overline{B} + C$

[7.44] (a) $A/(1 + AB)$ (b) 49.98 m (c) change of 2.36 m

[7.45] (a) 42.5 Ns/m (b) $s_{max} = 0.1177$ m/s (c) $s_{max} = 1$ m (d) 0.078 Hz (e) 0.75

[7.47] $T = 0.0125\ddot{\theta} + 7.5\dot{\theta}$ or $T = 0.0125\dot{\omega} + 7.5\omega$ newton metre; $f_n = 3.9$ Hz

[7.48] (a) 0.168 Hz (b) 22.77 Nms/rad (c) 0.145 Hz

[7.49] (a) $k = 4$ Nm/rad; $R = 2$ Nms/rad (b) 0.45 Hz (c) 0.707; 0.318 Hz (d) 0.4 Nm

[7.50] (a) $\ddot{\theta}_o + 2.91\dot{\theta}_o + 5.08\theta_o = 5.08\theta_i$ (assuming no tacho. F/B)
(b) 0.359 Hz; 0.646; 0.18 rad (c) 0.0786; 0.279 rad

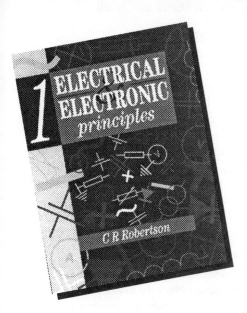

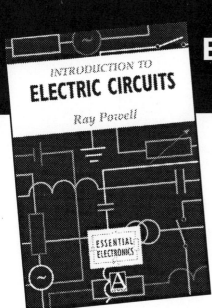